Astronomy is a science where the amateur and the professional interact constructively. This book is a testament to that relationship. *Images of the Universe* includes articles written by sixteen leading amateur and professional astronomers. They provide a lively and up-to-date view of our surroundings in space. The solar system, stars and galaxies all come under close scrutiny. Answers are provided to questions that have puzzled people throughout the ages, such as how and when was the universe created? Modern cosmology is discussed by Martin Rees and the first one second of the universe is put under the microscope by Paul Davies. Colin Ronan looks back on advances that astronomers have made in the last hundred years. Patrick Moore and David Hughes write on the outer planets, and comets and meteors respectively. Jacqueline Mitton examines the stars, the birth place of the elements. Heather Couper and Nigel Henbest turn their attention to the Milky Way, the hazy band of light that is the end-on view of our Galaxy; and Malcolm Longair looks beyond this and into the deep sky.

A good picture is the goal of most astronomers, whether it be beamed back to Earth by way of sophisticated and expensive technology, shot with a camera or simply drawn. In this book a special emphasis is placed on images of our universe produced by both amateurs and professionals.

This book will appeal to anyone interested in present-day knowledge of our surroundings in space. An awareness of our position in space may help the reader, but no specialist knowledge is essential.

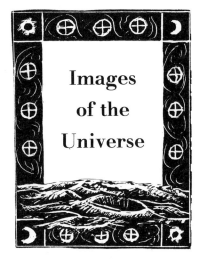

Images of the Universe

DATE DUE

WITHDRAWN

NOUMENA·TRANSCEND·PHENOMENA

A century ago the leading astronomers in the United Kingdom combined to form the British Astronomical Association. British astronomers have now joined together to produce this volume in celebration of the founding of the BAA in 1890.

Images
of the
Universe

EDITED BY

Carole Stott

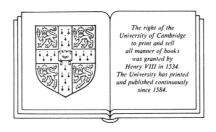

The right of the
University of Cambridge
to print and sell
all manner of books
was granted by
Henry VIII in 1534.
The University has printed
and published continuously
since 1584.

CAMBRIDGE UNIVERSITY PRESS

CAMBRIDGE

NEW YORK · PORT CHESTER · MELBOURNE · SYDNEY

Published by the Press Syndicate of the University of Cambridge
The Pitt Building, Trumpington Street, Cambridge C B2 1R P
40 West 20th Street, New York, N Y 10011-4211, U S A
10 Stamford Road, Oakleigh, Victoria 3166, Australia

First published 1991

Printed in Great Britain by
Butler & Tanner Ltd, Frome and London

A catalogue record for this book is available from the British Library

Library of Congress cataloguing in publication data available

Images of the universe/edited by Carole Stott
p.   cm.
Includs index.
ISBN 0 521 39178 4 (hb); ISBN 0 521 42419 4 (pb)
1. Astronomy.   I. Stott, Carole.
QB43.2.I43   1991
520—dc20   91–26321   CIP

ISBN 0 521 39178 4 hardback
ISBN 0 521 42419 4 paperback

Contents

Foreword

It is a pleasure to write a Foreword to this interesting and informative book, and particularly so in view of its being a tribute to a body that I greatly admire: the British Astronomical Association. The fact that the royalties will go towards promoting astronomy under the auspices of the BAA is an added bonus, and one with which I fully agree as I endeavour to improve the financial climate for the subject we all love.

The early history of the BAA makes fascinating reading, as those familiar with the 1890 edition of the *English Mechanic and World of Science* will know. There, one reads of the problems of the BAA's predecessor, the Liverpool Astronomical Society, and of the activities of such authors as E. Walter Maunder and William Huggins, and the Reverend T. E. Espin, the latter moving to a parish not far from Durham where he established a fine reputation in a variety of fields. One also notices that E. E. Baley, Deputy Chief Cashier of the Bank of England was nominated as a scrutineer for the elections for the Council. There is food for thought here!

Turning to the role of amateurs in contemporary astronomy, this is considerable in so many ways. Planetary science could not do without them – for studies of comets, meteor streams and meteorites, and other observations, too. Equally important is the network of talented people eager and willing to disseminate the results of the professionals both in the UK and abroad. In this the BAA plays a most important part. The interaction of the amateur and the professional in astronomy is probably stronger than in any other science and to the benefit of both, and citizens as a whole.

I know that the authors of these articles have gained much pleasure from writing them; may the readers gain even more, and may their number be astronomical!

Professor Arnold Wolfendale, FRS
Astronomer Royal

1

Major advances in astronomy since 1890

COLIN RONAN

Over the last one hundred years, astronomy has seen an advance unprecedented during any previous century. This has been due primarily to two factors. In the first place there has been a vast growth in ways of observing the universe. Not only have telescope apertures increased in size, allowing ever dimmer objects to be discerned and so enabling astronomers to penetrate much deeper into space, but also totally new techniques have made it possible to examine hitherto unavailable evidence. For the first time in history, astronomers are now able to study all radiation emanating from the universe, no longer restricting themselves to one narrow range of the spectrum − visible light. Space-craft and electronic techniques have also made it practicable to probe the planets of the solar system in ways still thought impossible even thirty-five years ago.

The second significant change which has come about in astronomy is due to the increasingly effective interplay between astronomy and physics − particularly particle physics. Admittedly this last began more than a century ago, but only within the last hundred years has it been possible for these results to yield their true significance. In consequence, astronomers now understand the basic processes by which stars form, generate their energy, and age or evolve. They are able not only to show that some of what they once called nebulae are in fact external galactic systems of which ours is only one of millions, but also that all are moving away from each other − in brief that we live in an expanding universe. They have discovered exotic new objects such as pulsars and

1. Reflecting telescopes could be built with larger apertures than refractors. The 100-inch reflector at the Mount Wilson Observatory completed in 1917 was the largest in the world for some decades.

quasars, and are even able to apply the theory of relativity and the quantum theory of matter to discuss scientifically the origin of the universe and its ultimate fate.

Yet even as early as 1897 the fundamental importance of this marriage between astronomy and physics was appreciated by George Ellery Hale, that indefatigable progenitor of large aperture telescopes, when he remarked that the new Yerkes Observatory with its vast 40-inch (1-metre) aperture refractor, was 'in reality a large physical laboratory as well as an astronomical establishment . . .'

Although the Yerkes telescope was a refractor, Hale was the central figure during the past century to promote the move to reflecting telescopes because these could be made with apertures far larger than any refractor. Mounted with all the aplomb of twentieth-century mechanical engineering, such telescopes have revolutionised observations from Earth-based observatories.

This takeover by the reflector began in earnest in 1908. Ever since then apertures have kept increasing, so that now a 10-metre aperture reflector, the Keck Telescope, is to be established on Mauna Kea in Hawaii. This instrument will be another of a new generation of optical telescopes which will make use of computer-control of the optics in a novel way, devised originally in the late 1970s for the United Kingdom Infra-Red Telescope (UKIRT) in Hawaii. In the UKIRT instrument the mirror is much thinner than usual — a factor which reduced the cost both of the mirror itself and its mounting — and is supported all over its rear surface by pads, each hydraulically controlled so that the shape of the

2. The Mauna Kea Observatory in Hawaii, home to some of the world's largest telescopes. The Keck telescope, a 10-metre aperture reflector, is the most recent addition.

mirror does not alter as the telescope moves to observe different celestial objects. This method allows a precision of control which the old mechanical linkage supports of the past could never attain. The New Technology Telescope (NTT) of the European Southern Observatory in Chile finished in 1989 has adopted this by computer control techniques with notable success at optical wavelengths which, being much shorter than infra-red, impose more stringent requirements. Because a solid 10-metre diameter mirror would be impossible to construct, the technique is being taken a stage further in the Keck telescope, where the 10-metre mirror is in fact a collection of thirty-six small hexagonal mirrors, all of which the computer-controlled supports will retain in perfect optical alignment.

Another radical change to telescopes since the establishment of the reflector came in 1932 when the Estonian Bernhard Schmidt announced a totally new design. Schmidt's aim was to devise an instrument with a wide field of view of the order of seven degrees across, compared with the maximum of about one-fifteenth of a degree of the then largest astronomical telescopes. He succeeded by inventing a catadioptric system — a cross between a refractor and a reflector — though his telescope could not be used visually as its image was accessible only to a photographic plate placed inside the instrument. Photographic surveys of large areas of the sky are now commonplace, making possible observations which otherwise would have been totally impracticable.

Optical observing has also been much affected by the use of electronics and computers.

Light-sensitive devices are now in wide use. Among them are electronic cameras with photo-cathodes which multiply the electrons emitted by their light-sensitive surfaces and record them on photographic plates, all essentially developments of a design originated by the French astrophysicist André Lallemand and used first in 1947–8. Yet even higher sensitivity is attained by the Image-Photon-Counting System. Invented in the 1980s by the British astronomer Alec Boksenberg, this uses an image-tube to intensify the light received by a telescope some 700 000 times, and then processes the output with a computer, building up an image which can be studied visually. Another product of the same decade is the development within the electronics industry of the charge-coupled-device (CCD), a silicon chip divided into some 640 000 separate sensitive areas which feed their results directly to a computer, Such developments as these have meant an increased sensitivity for earthbound telescopes, whatever their size. All the same, photographic techniques still play a significant part; they continue to be an important way of recording information, while increased sensitivity of emulsions and, in the last decade, the use of 'unsharp masking techniques' have made it possible to detect material previously inaccessible to the photographic plate, as the recent discovery (1989) of a giant galaxy Malin 1 (named after the photographer David Malin) bears witness.

There has also been considerable development of that basic attachment to the telescope, the spectroscope. So that the lines in a spectrum may be analysed and measured more precisely to give greater understanding about the physical processes within stars, and to measure 'line-of-sight' or radial velocities with increased accuracy, new designs of spectroscope have been developed. Multi-slit instruments, spectroscopes using interference of the incoming light to form a spectrum, and either novel 'objective prisms' for fitting at the front end of a telescope, or special direct-viewing instruments such as that designed by Patrick Treanor whereby the spectra of a host of stars may be photographed at a time, have been developed. These allow spectral surveys to be made and also great numbers of radial velocities, particularly of high velocity objects such as distant galaxies and quasars, to be measured. Image dissectors using fibre optics have even been introduced to increase precision measurement, and both photon counters and CCDs are used so that spectra of much dimmer objects can now be measured.

In the last twenty-five years, other applications of electronics and computer assistance in the optical field have been in the processing of observations and in the increasing automation of telescopes themselves.

As far as the processing of observations is concerned, there are two aspects. One is the examination of photographic plates. Even twenty years ago astronomers had the laborious task of measuring with a special microscope each separate celestial object recorded on the plate. This could be, and was, extremely time-consuming; with plates taken using Schmidt telescopes the vast amount of information available became virtually impossible to handle. At last, in 1969, a great measure of automation came with the invention of a machine which measured the brightness of each separate image and its position on the plate. A commercial development, made in conjunction with Vincent Reddish of the Royal Observatory at Edinburgh, it has made possible the measurement of at least 900 images per hour. Similar devices have been built elsewhere, notably an automatic instrument, developed by Edward Kibblewhite at Cambridge University in the

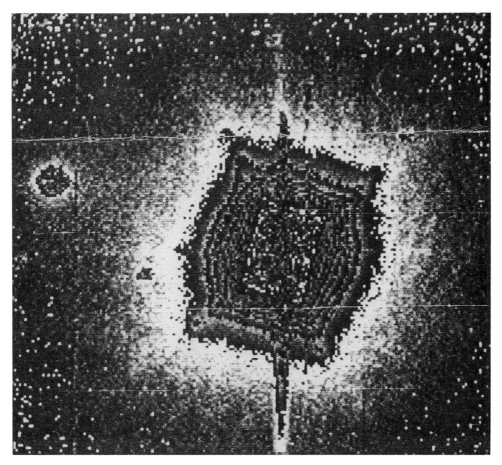

3. This CCD (charge-coupled-device) image was obtained using the Anglo-Australian Telescope. The rectangle is a region of very faint nebulosity surrounding a bright star.

1970s, which will not only measure images but also select those given by objects of different kinds – galaxies, stars, etc. – and chart the results. However, measurements using computer programs alone are now possible if the output from CCDs is recorded on magnetic tape.

The second technique which has been developed by many astronomers is the computer-processing of observations. Here specialised computer programs provide images with the various aspects under investigation being emphasised by display in different colours. Images of this kind are now produced not only for optical observations but also for those made at other, non-visible, wavelengths. They can be seen throughout the rest of this book.

As far as telescope automation is concerned, in 1948 the 200-inch reflector had a computer to correct the setting of the instrument on any selected celestial object for apparent distortions of position due to effects of the Earth's atmosphere. But since the 1960s complete automation has been achieved, telescopes being designed so that they automatically follow selected celestial objects on which they can set themselves once the coordinates are programmed into their computers. This not only relieves astronomers of tedious sessions at the telescope, but also speeds up the amount of work which can be done each night with a particular instrument and, at the same time, allows telescopes to be operated by remote control over intercontinental distances. But of course the most sophisticated of all remote control applications is that applied to orbiting space-craft such as planetary and lunar probes, X-ray satellites such as the American Einstein probe of 1978 and the European Space Agency's Exosat launched in 1983, and the infra-red astronomical satellite IRAS, also placed in orbit in 1983.

Mention of the European Space Agency underlines another new factor in astronomy of the twentieth century, namely the growth of international cooperation, which is coupled with the fact that astronomers now tend to work in teams rather than as isolated individuals. International cooperation has also been a feature of this century's astronomical scene. In 1922 the International Astronomical Union was formed to encourage cooperative planning of observational programmes and the adoption of internationally agreed constants, constellation boundaries, and so on. It also promotes specialist symposia in various countries.

The use of computerised control has also meant that, since the late 1950s, telescopes could be mounted on the simpler altazimuth mounting instead of the expensively engineered equatorial with its axis tilted over at an angle depending on the latitude of the observatory. The computer does the necessary calculations for following a celestial object as it appears to move across the sky. Although this has only come into general use for optical telescopes within the last decade, it now means that costs of construction can be reduced and that very large telescopes, like the Keck 10-metre, become a practical possibility.

However, the computer-controlled altazimuth mounting was first designed for use with radio telescopes, the first of the non-visual wavelength instruments with which astronomers have begun to observe the universe since the Second World War. Admittedly, attempts to observe radio signals from the Sun were made between 1894 and 1902, but all proved unsuccessful since the equipment used was unsuitable, and the true beginnings of radio astronomy began serendipitously in the United States in 1930 when Karl Jansky was making studies of radio 'static and interference'. He noted that some of the static appeared to come from the region of the Milky Way. However, astronomers appear to have taken but slight interest in his observations, which were followed up only by Gert Reber, an amateur radio enthusiast, who in 1937 built the first specially designed radio telescope, a dish 9.4 metres (31 feet) in diameter. Between 1940 and 1942 Reber published more detailed observations than Jansky had been able to obtain, and in 1946, after wartime hostilities had ceased, his results stimulated others to take up the work.

In the United States emission from the Sun was detected by Reber and George Southworth, and research was pursued at the Naval Research Laboratory at Georgetown University, but it was in Britain that James Hey, making use of wartime radar equipment, discovered in 1942 radio emission from the Sun, and in 1945 used radar itself to detect meteors. His efforts stimulated a number of radio physicists in Britain and Australia to pursue radio astronomical research; in particular they led Martin Ryle to begin establishing

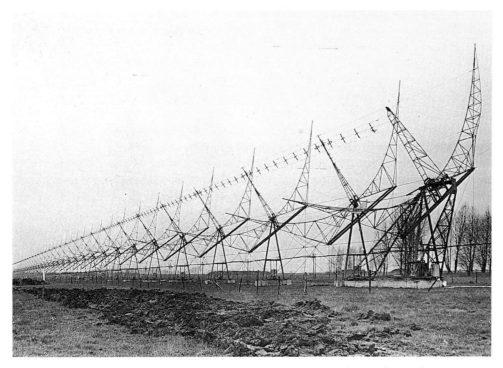

4. *In the early 1940s radio physicists turned to radio astronomical research. A radio astronomy observatory was established at Cambridge in 1945 and telescopes were built to survey the sky for radio sources.*

a radio astronomy observatory at the University of Cambridge in 1945, at almost the same time as Bernard Lovell at the University of Manchester. In Australia, Edward Bowen and Joseph Pawsey of the Commonwealth Scientific and Research Organization set up a radio astronomy group, also in 1945, while in 1946 groups formed in both Canada and in The Netherlands.

Since radio wavelengths are at least some 100 000 times longer than those of light, to begin with radio astronomers were unable to pinpoint celestial radio sources precisely enough to incorporate their results fully into the main body of astronomical knowledge. To overcome this serious disadvantage, therefore, interferometers for radio observation were developed very early on in both Australia and Cambridge. However, interferometers in astronomy were nothing new. In 1919 Albert Michelson had fitted such an instrument of his own design to the 100-inch reflector in order to measure stellar diameters by causing the light waves from two beams of light from a star to interfere, and successful measures of Betelgeuse were made the following year. Bowen and Pawsey's group used different designs for radio astronomy, but Ryle developed Michelson's method, which, in the late 1950s, was to yield the powerful technique of aperture synthesis using a collection of electronically linked radio telescopes, the interfering signals from which were analysed by computer. From 1967 onwards techniques for electronically coupling radio telescopes in different countries to give giant interferometers have become possible, so that radio astronomy has been able to compete and even surpass optical astronomy in discerning detail in celestial objects observed by both techniques.

The success of radio astronomy in discovering neutral hydrogen gas, which radiates no visible light (1951), objects like quasars (1963) and pulsars (1967), and even complex molecules in space (1968 onwards), alerted astronomers to the advantages of probing space in non-optical wavelengths. Although heat or 'infra-red' rays were discovered in 1800, their significance in astronomy remained generally unappreciated until the 1960s. Only then did sensitive detectors for such radiation became available and astronomers begin to appreciate that this radiation band could help give insight into regions both within and outside our own Galaxy, which were blacked out optically by clouds of dust. Since then, infra-red detectors from Earth-based equipment such as the United Kingdom Infrared Telescope (UKIRT) of 1979 and from space-craft like the 1983 Infra-red Astronomical Satellite (IRAS) have made notable contributions to astronomy, such as the detection of dust rings round stars, the discovery of very low mass stars – 'brown dwarfs' – and have shed much light on the nature of clouds of gas molecules in those regions where stars are formed and, indeed, on the nature of star formation itself.

The short end of the wavelength range – ultra-violet emissions, X-rays and gamma rays – have also produced new evidence for astronomy. Again, although ultra-violet radiation was discovered in 1801, and X-rays were detected first in 1895, little astronomical use was made of the former and none of the latter until it became practicable to make observations from high up or even outside the Earth's atmosphere. High-flying aircraft and then high-altitude balloons were used first for such observations in the late 1940s, but it was when rockets were sent up in the 1950s that studies in X-ray and the far ultra-violet wavelengths began to gather momentum.

After the launch by the Russians of the Sputnik space-craft in 1957, the practicability of using artificial orbiting satellites became realised, and in 1962 an exploratory study led by Riccardo Giacconi in the United States brought about the discovery of an intense X-ray source, later to be called Scorpius X-1; this appears to be a binary system in which one star is drawing away material from the other, which seems to be a black hole. Since then X-ray probes like Uhuru (1970), Einstein (1979) and Exosat (1983) have shown up the existence of slowly rotating neutron stars and even intense X-ray sources within some galaxies such as Centaurus A. They have also shown the presence of otherwise invisible material around some galaxies, and even of gas at the very high temperature of 100 million degrees lying right within clusters of galaxies.

The full ultra-violet range of wavelengths became available for study from the early 1970s onwards, and most notably with the International Ultraviolet Explorer (IUE) of 1978. Used primarily to examine sources using spectroscopes, ultra-violet probes have already provided much detailed information on the gas lying between stars – the interstellar medium – and the nature of very hot stars which radiate very strongly at ultra-violet wavelengths, showing up the existence of stellar winds from such stars and the transfer of material from one star to another in some binary star systems. Moreover, spectroscopic studies of the ultra-violet radiation from some nearby galaxies have shown certain interesting similarities between them and quasars.

As far as cosmic rays are concerned, their study has a long history which goes back at least to 1910 when Theodor Wulf took instruments to the top of the Eiffel Tower in Paris to determine how discharges of electrical equipment, first discovered in the laboratory, varied as one rose above the Earth's surface. He found that the discharges increased at a height of 300 metres (984 feet), and this suggested an extraterrestrial source for the radiation causing them. Confirmed on balloon flights taken the next year by Victor Hess, and in 1913 by Werner Kohlhörster, they were named 'cosmic rays' by Robert Millikan in 1925. Subsequent investigations during the 1930s by Patrick Blackett at Manchester and Cecil Powell at Bristol showed that very short wavelength gamma rays were the discharging agency, but that these were produced by atomic particles arriving from space. From the late 1940s onwards studies have increased, and by the 1980s it had become evident that not only is the Sun a source of such radiation, but so also is the Galaxy, and that such radiation is emitted from other galaxies as well. The atomic particles are primarily protons; they possess immense energies and, at least for the most part, come from the remnants of supernova explosions.

For the greater part of the last century, observation of the Moon and the planets of the solar system has been primarily the province of the amateur observer, though the first satisfactory photographic atlas of the Moon was made by the professionals Maurice Loewy and P. H. Puiseux between 1896 and 1910. Nevertheless, amateurs made far more detailed studies of the lunar surface and of the planets than photography could achieve; indeed it was Percival Lowell, a business-man turned astronomer, who was the central figure in the great controversy about the martian surface, which began in 1895 with Giovanni Schiaparelli's description of *canali*. Schiaparelli was adamant that these should not be regarded as watercourses, but Lowell in particular believed they were and charted them with care during the succeeding decades. Only now, with the advent of space probes, has the matter been settled and the canals found to be a very persuasive optical illusion.

However, in the last thirty years, observation of the solar system has been revolutionised by the use of space-craft. The far side of the Moon, invisible to Earth-based observers, was photographed in 1959 and again in 1965 by Russian space-craft, and American probes took close-up photographs of the lunar surface from 1964 onwards. In 1970 and 1973 the Russians landed remote-controlled unmanned vehicles on the lunar surface, but already by 1969 the Americans had achieved manned landings and had managed to place experimental equipment on the Moon for later use; they also brought back to Earth samples of lunar rock, the first pieces of the surface of another celestial body to be artificially recovered for later study.

As far as the major planets are concerned, Venus seems by and large to have become the province of Russian probes, which started exploration in the 1970s, though American craft mapped the surface by radar at the end of 1978. For the rest, American space vehicles have taken very good close-up photographs of Mercury (1974), Mars (1964 to 1975), Jupiter (1972 to 1979), Saturn (1980 and 1981), as well as Uranus (1986) and Neptune (1989), including some detailed images of their satellites, and have achieved two soft landings on Mars in 1976. These last have analysed martian soil and led to the conclusion that, whatever might have been the case in the distant past, Mars harbours no living

Major advances in astronomy since 1890

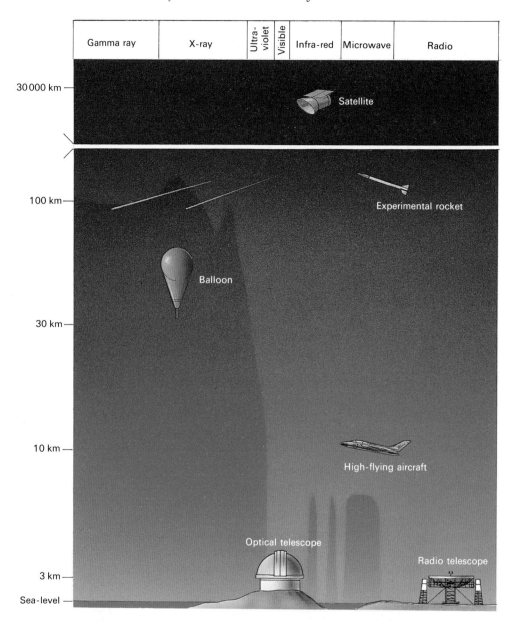

Gamma ray	X-ray	Ultra-violet	Visible	Infra-red	Microwave	Radio

30 000 km — Satellite

100 km— Experimental rocket

Balloon

30 km—

10 km — High-flying aircraft

Optical telescope

Radio telescope

3 km—

Sea-level —

5. *Astronomers now study all the different forms of electromagnetic radiation given off by astronomical objects. But not all wavelengths reach Earth, as indicated by the shading. Specialised instrumentation and techniques have been developed to collect and investigate the different wavelengths.*

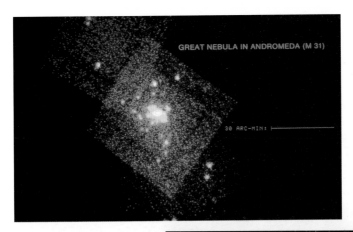

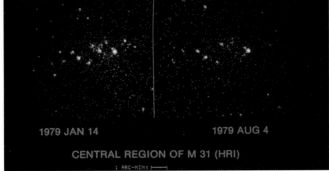

6. The Einstein Observatory satellite was used from 1978 to 1981 to make X-ray observations of the sky. These are X-ray views of a spiral galaxy, the Andromeda 'Nebula' (M31), the most distant object visible to the naked eye.

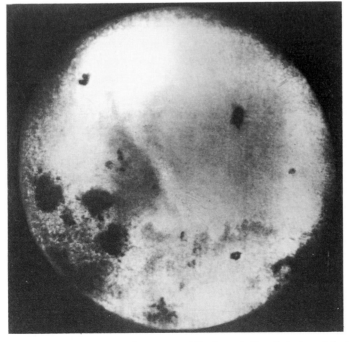

7. The same face of the Moon is always turned Earthward. Our first view of the far side was obtained by the Russian probe Luna 3 in October 1959.

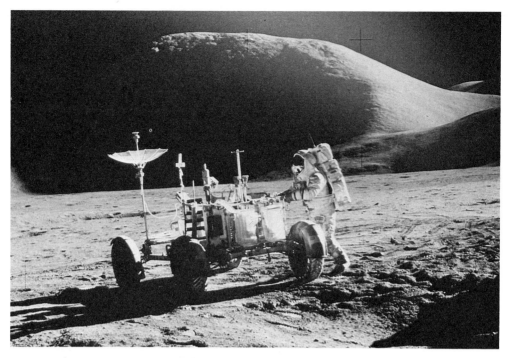

8. *American astronauts travelled across almost 90 km of the Moon's surface on the six Moon landings between 1969 and 1972. Here James Benson Irwin, the eighth man to walk on the Moon, is seen with the Lunar Rover.*

material of any kind. Space-craft have also made it possible to confirm in 1986 that Fred Whipple's theoretical model of a comet as a ball of icy material, frozen gases and dust devised in 1950 is correct, at least as far as Halley's comet is concerned, and, as a whole, the results of probes within the solar system have given great impetus to professional lunar and planetary studies, which had lapsed almost completely after the Second World War.

Dynamical studies of the solar system coupled with the discovery of many minor planets – between 1891 and 1901 Max Wolf used photography to detect more than 200 – allowed the determination of the distance from the Earth to the Sun to be obtained with hitherto unprecedented accuracy when Eros came close to the Earth in 1900 and 1931. Nevertheless, radar echoes from Venus obtained between 1961 and 1962 have superseded even those observations. Another notable dynamical achievement was the discovery of Pluto by Clyde Tombaugh in 1930, based on calculations by Lowell and William Pickering, although serious doubt is now being cast on whether Pluto was the disturbing force on Neptune on which the calculations were based.

In spite of these spectacular planetary results, studies in astronomy over the last century have been primarily concerned with the stars and other objects deep in space such as nebulae. These last have changed completely the face of astronomy and have brought about a view of the universe which is far grander and more astounding than ever

9. In 1979 Jupiter was visited by both Voyager 1 and 2. The two space-craft took more than 33 000 pictures of the planet and its five major satellites. This image shows the neighbourhood of Jupiter's Great Red Spot.

previously conceived. When 1920 dawned, the Milky Way was thought to mark the boundaries of the universe, yet within a decade it became evident that it was only one island of stars, dust and gas among a myriad of others, extending as far as telescopes could see. This revolution was based on much painstaking study of the physical nature and distribution of stars of every kind.

The examination of orbiting double stars or binaries has shown their immense number; it is estimated that perhaps more than half the stars within the confines of the Milky Way are binaries or multiple stars. The great number of binaries, catalogued first in 1906 by Sherburne Burnham, has important implications for the process of star formation.

Variable stars have also received intense study during the last century, due not least to increasingly accurate methods of brightness measurement. Most significantly, this led Miss Henrietta Leavitt in 1912 to recognise that Cepheid variables – those variables named after Delta Cephei – had periods of variation which depend on their true brightness. This gave astronomers a new yardstick for determining distances in space far beyond those which could be obtained by direct measurement. The implications of this have also had immense significance on ideas about stellar formation.

Another type of variable, the nova − a star too dim to be visible ordinarily which becomes so bright as to be readily observed and so appears as a new object in the sky − has received much attention. Known in the West from 1572, such stars were classified into 'novae' and 'supernovae' by Walter Baade and Fritz Zwicky in 1934, and the latter have been the subject of intense study, since their brightness increases for only a week or so by a factor of some 10 000 million times. In 1941 Rudolph Minkowski classified supernovae into two types, while more recent supernova research has led astronomers to much deeper understanding of stellar evolution.

The study of stellar spectra has brought about greater insight into the nature of stars themselves, and, coupled with modern particle physics, has made it possible for sources of stellar energy to be understood in some detail. Although the first steps were taken in 1867, it was only in 1890 that there appeared a catalogue of spectra, named after Henry Draper who had initiated the research leading to its compilation. This Draper classification, duly modified by Miss Annie Cannon, appeared between 1918 and 1924 and still forms the basis of today's system.

Analyses of spectra and the colours of stars, together with a study of their true brightness, were made separately by two astrophysicists, Henry Norris Russell and Ejnar Hertzsprung. These resulted in Hertzsprung producing in 1911 diagrams comparing colour and brightness for stars in the Hyades and Pleiades clusters, and in 1913 with Russell publishing a diagram of brightness and spectra for a wide variety of stars. This last gave rise to what has become known since as the Hertzsprung–Russell or 'HR' diagram. Displaying graphically vital stellar characteristics, the diagram stimulated much subsequent research into stellar evolution.

As early as 1877, Russell had taken up the subject of the evolution of stars, and by 1910 he was convinced that they began as cool red bodies of great brightness − red giants − grew hotter as time passed, and then shrunk to become small dim red objects − red dwarfs. But as theoretical research into the internal constitution of stars and the means by which they generated their radiation progressed during the years that followed, Russell's sequence had to be abandoned.

It was Arthur Eddington who by 1917 first appreciated the vital role which radiation played in preventing a star collapsing under its own gravity, though his views met with opposition. However, by 1924, Eddington had discovered the crucial relationship between the mass of a star and its brightness − his mass/luminosity law − and the 1920s also saw contributions from Celia Payne on the relative abundances of chemical elements in stars and by Russell on the composition of the Sun.

During the previous decade, in 1915, Walter Adams had obtained spectra of dim white stars, notably that of the companion of Sirius. These discoveries of white dwarfs made it clear that some stars could have densities far greater than those obtainable from ordinary matter. This led Eddington to suggest that in stellar interiors atoms must be stripped of their electrons, allowing atomic nuclei to be packed unusually closely together.

As to the generation of stellar energy, this had long presented problems. By the 1890s the age of the Earth was known to be very great, and after the discoveries of the

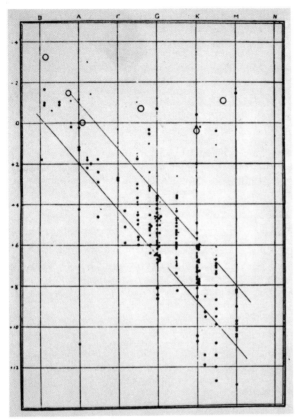

10. Russell's first diagram, which gave rise to the Hertzsprung–Russell diagram relating a star's intensity to its spectral type.

radioactive breakdown of atoms in 1896 by Henri Becquerel, and of radium in 1898 by Marie and Pierre Curie with the assistance of Gustave Bémont, it became clear that ages of the Earth's rocks must be counted in thousands of millions of years. However, no chemical or physical mechanism was then known to account for the generation of the Sun's energy over so long a period. Only in 1907, when Einstein showed that mass and energy were equivalent – symbolised in the now famous equation $E = mc^2$ – was a new road open, although it was not until after Eddington's work just mentioned that the synthesis of atoms inside the Sun could be appreciated as the probable explanation. In 1931 Robert Atkinson suggested this, and was followed up by Carl von Weizsäcker in 1938 and Hans Bethe in 1939. In that same year Ernst Öpik put the explanation into the context of stellar evolution, thus paving the way for studies which culminated in 1957 with the work of Geoffrey and Margaret Burbidge, William Fowler and Fred Hoyle on the synthesis of chemical elements inside stars. This proved to be of immense significance in the years that followed for a deeper understanding of stellar evolution and the composition of the cosmos.

Long before 1890, globular clusters of stars and the spiral structure of some nebulae had been detected. But it was not until 1908 that both spectroscopy and photography showed that the spiral nebulae seemed to be composed of stars, and also that Cepheid variables existed in globular clusters, thus giving an indication of their distances. Again,

11. *These photographs of a spiral arm of the Andromeda Nebula, and its companion galaxy NGC 205, were taken using the 200-inch Hale telescope. Walter Baade's 1944 photographs of the Andromeda Nebula, taken with the 100-inch, were the first to show that it was made up of separate stars.*

though it was known that the Sun was moving in space, it was not until measurements of the motions of other stars had been made, particularly by Lewis Boss from the 1890s onwards, that these two fields of observation could begin in the 1920s to lead finally to a complete revision of ideas about the universe.

Nevertheless, as early as 1904 the Dutch astronomer Jacobüs Kapteyn detected what appeared to be a tendency for stars to move across the sky in two streams in opposite directions. In 1913 Eddington suggested that this was due to the rotation of the whole star system of which our Sun is a member; in 1925, Bertil Lindblad examined all the evidence in detail and came to the same conclusion. The star system Lindblad considered was that proposed by Harlow Shapley, who in measuring the distances of globular clusters had concluded in 1915 that it was of vast size, with a diameter of no less than 300 000 light-years. But not all astronomers accepted Shapley's estimates, nor did they accept his belief that the spiral nebulae lay inside his star system. Many believed that they were external 'island universes'.

In 1925 the question was finally resolved when Edwin Hubble announced the discovery of Cepheid variable stars in two spirals, for his results placed them at distances of over 900 000 light-years. Visual confirmation came from Walter Baade in 1944 at Mount Wilson, who took some astounding photographs which showed separate stars in the Andromeda 'nebula'.

But if the spiral nebulae – or galaxies as they have been called since the 1950s – are collections of stars, dust and gas, they present another feature of immense significance in this century's understanding of the universe. In 1917 Vesto Slipher found that almost all spirals showed prodigious red-shifts in the spectra, shifts which were known to indicate that they are moving away from us. This became of immense significance when, in 1929, Hubble announced evidence which unequivocally confirmed a suggestion made in 1921 by Carl Wirtz that the more distant a galaxy is, the faster it moves away from us. Then, in 1934, Hubble was in a position to formulate a velocity–distance relation which has since become known as Hubble's law, although over the years its numerical value has been modified in the light of later evidence; its basic principle is not, however, in question. We live in an expanding universe.

Observations amassed during the last century have brought into prominence as never before the question of the origin and future of the universe itself. Mathematical cosmology has become a rich field of development, yet this is due not only to stimulation from observational astronomers, but also to two crucial twentieth-century theories – relativity and quantum theory.

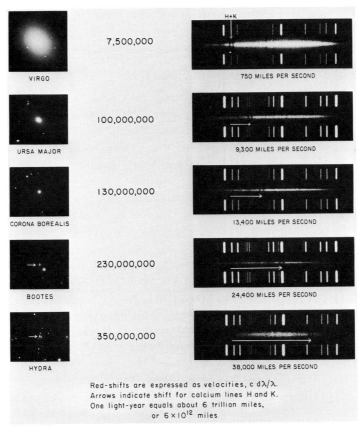

Red-shifts are expressed as velocities, c dλ/λ.
Arrows indicate shift for calcium lines H and K.
One light-year equals about 6 trillion miles,
or 6×10^{12} miles

12. *Edwin Hubble's montage of five extragalactic 'nebulae' (galaxies), in order of increasing distance, with their red-shifts. The more distant the object the faster it is moving away from us.*

It was in 1905 that Albert Einstein published his 'special' theory of relativity, 'special' in that it dealt only with observers and 'frames of reference' moving with constant velocity in respect to each other. All such frames were equivalent, none had preference over the others. In addition, in accordance with work by some other theoretical physicists, the velocity of light was taken to be the maximum velocity of any body in the universe and to be the same in every reference frame. In the following decade Einstein continued this work, and in 1915 he was able to publish his 'general' theory of relativity; here the frames of reference could be also in relative accelerated motion. General relativity is essentially a theory of gravitation, and one more accurate in expressing the behaviour of the universe than Newton's. Astronomically it accounts for a hitherto unexplained discrepancy in the motion of Mercury's orbit, and for the deflection of starlight observed during a total solar eclipse. But there are many other consequences in astrophysics, not least the behaviour of black holes, while most particularly it has helped bring deeper understanding to ideas about the beginning, and end, of the universe.

Such ideas have also been revolutionised by the application of quantum theory, which developed after Max Planck had in 1900 introduced the concept that energy is emitted in discrete amounts or 'quanta'. The results of the theory are a powerful means of explaining white dwarfs, neutron stars and other 'degenerate' matter, as well as allowing cosmologists in the last few decades to trace back, in conjunction with relativity, the early history of the universe.

From the 1920s to the 1960s these ideas stimulated two basic theories of cosmology. In 1927 Georges Lemaître suggested that the universe began with a 'primeval atom' or 'singularity' from which it expanded to its present state. However, later studies by Hubble on the rate of expansion, and by others on the age of the Earth, made it clear that there seemed to be a grave discrepancy – Lemaître's theory showed that the universe appeared younger than the Earth itself. In consequence, in 1948 Hermann Bondi and Thomas Gold proposed a 'steady-state' universe which had existed for ever, and which underwent no change on the largest scale; loss of density as galaxies move outwards was made up for by the continuous creation of fresh matter throughout the whole of space. The same year Fred Hoyle showed that such a universe could be derived from Einstein's relativity equations.

In 1952, however, revisions of the Hubble expansion rate by Walter Baade brought a greater age for the universe and removed the need for formulating the 'steady-state' theory. Yet only in 1965, when Arno Penzias and Robert Wilson detected the existence of a background of radiation all over space, was it discarded. This was because their discovery brought universal realisation that such radiation was firm evidence for a 'hot' big bang beginning to the universe, as proposed first in 1935 by George Gamow and then in more detail by Gamow, Ralph Alpher and Hans Bethe in 1948.

With relativity theory and the results of particle physics, together with corroboration from recent observations, astronomers are now of the opinion that they have at least a general understanding of the origin of the universe. In conjunction with nuclear physicists, cosmologists are now working out the actual processes by which a big bang could give rise to such a universe as we observe, and feel some confidence in their general interpretation of events back as far as one-ten million billion billion billion billionth seconds (10^{-43}s) after the big bang itself. They are even seeking the unification of gravity, nuclear forces and electromagnetism into one grand unified theory. However, there are difficulties at present, both in complete unification and in getting back to earlier fractions of a second; even Einsteinian gravitation breaks down and at least a new theory of gravity is needed; this is actively being sought.

The ultimate future of the universe has also attracted much attention; will it expand for ever or begin to contract? Present evidence is somewhat ambivalent. Perhaps, in the event, observations by space-probes will bring such a novel slant on what we know that more than simple revision of what seem to be well established views may be required. But it seems unlikely that much of the knowledge about the nature and evolution of stars, the distribution of galaxies, and the expansion and general layout of the universe gained during the last hundred years will be proved wrong, although the history of astronomy shows that one should always be wary of being too certain. There is always the possibility that during the next century astronomers will be forced to revise some currently held opinions and expand their horizons even further, as has happened time and again in the past.

2

The inner planets

RICHARD BAUM

Of the solar system's nine primary planets, the four nearest to the Sun are known as the inner or terrestrial planets. They are, in order of distance from the Sun, Mercury, Venus, Earth and Mars. All move in orbits within 1.52 A U of the Sun, and are similar in size and density. All have rocky surfaces and can be studied by geologic techniques. Unlike the jovian or outer planets, the terrestrial planets have few, if any, satellites. Mercury and Venus have none; Earth has one, the Moon; and Mars has two small ones, Deimos and Phobos.

Until December 1962, when Mariner 2 flew past Venus, our knowledge of the terrestrial planets, other than the Earth, was very limited. Centuries of direct visual observation had established a sound understanding of their orbital theory, sizes and densities, and the rotation of Mars, but virtually nothing about their surfaces. Mercury was small and too near the Sun for effective observation. Venus hid its secrets beneath a thick blanket of cloud; while our thinking about Mars still reflected the ideas of Percival Lowell. What follows is a brief exposition of these worlds as they are perceived in the aftermath of the first phase of space exploration.

Mercury

The innermost of the planets, Mercury, dashes around the Sun once every 87.97 days at a distance which varies between 0.3074 A U (perihelion) and 0.4667 A U (aphelion) in an orbit inclined at 7° to the ecliptic. It darts into the twilight, sparkles briefly, and as quickly is gone, its performance largely unregarded.

Blinding proximity to the Sun invests Mercury with a certain elusiveness, and renders it extremely difficult to observe from the surface of the Earth. At maximum elongation it is never more than 28° from the Sun, and is only observable in daylight and bright twilight. But in daylight the brightness of the Earth's sky reduces contrast on the tiny disc, while at twilight the planet is low in the sky, and its image is then distorted by the murky, turbulent air near the horizon. Viewed less as a world than as a probe for very sensitive tests of gravitational theory, Mercury accordingly failed to attract much interest prior to the Mariner 10 reconnaissance. What data we did have related to its physical

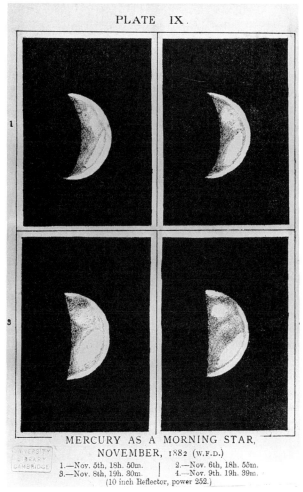

PLATE IX.

MERCURY AS A MORNING STAR,
NOVEMBER, 1882 (W.F.D.)

1.—Nov. 5th, 18h. 50m. | 2.—Nov. 6th, 18h. 55m.
3.—Nov. 8th, 19h. 30m. | 4.—Nov. 9th, 19h. 39m.
(10 inch Reflector, power 252.)

1. Mercury drawn by W. F. Denning in November 1882 from observations made using a 10-inch reflector. Mercury is shown for 5 and 6 November (top left and right) and 8 and 9 November (bottom left and right).

characteristics and orbital theory. Scanty as it was, this told of a curious, puzzling little world, best described as a thin-shelled iron sphere.

Of course, a few early telescopists were successful, and Giovanni Zupus (c.1590–c.1650) in particular is cited for his discovery of the phases in 1639. Nevertheless, other than the phase all the telescope usually showed was a small fuzzy spot of dull, amber light, broken by indistinct greyish patches and an occasional white spot, from which a rotation of twenty-four hours was initially deduced. But that owed more to terrestrial analogy than scientific objectivity, and in 1890 was supplanted by Giovanni Schiaparelli when he revolutionised our view of the planet by announcing it was locked in synchronous rotation. In other words, like the Moon, Mercury spun on its axis in precisely the same time as it took to circle the Sun. Unlike the Moon, however, one hemisphere baked in perpetual day, the other froze in eternal night. It was a convincing portrait that became a cliché. Yet like many of our cherished assumptions about the planets, it was shattered by space age technology.

Suspicions were first aroused in 1962 with the unexpected detection of radio emission from the supposedly frigid night side. Within three years the mystery was solved. Radar data showed that Mercury actually spins on its axis in 59 ± 5 days, later refined to 58.646 ± 0.005 days. Coupled to the 88-day orbital period, this gives Mercury a solar day of 176 Earth-days, since we now know the planet spins three times on its axis during two complete revolutions around the Sun.

This 3:2 resonance, or spin–orbit coupling as it is known, is a more complex example of that found in the Moon, and has peculiar consequences for the motion of the Sun across Mercury's sky. Its most important effect is that at alternate perihelion passages the same hemisphere always faces the Sun. In tandem with the planet's high orbital eccentricity, this results in unequal heating of the surface. Hence, the subsolar points at longitudes 0° and 180° are referred to as 'hot poles', because they receive optimum radiation at perihelion, while the subsolar points at 90° and 270° are known as 'warm poles', since they are illuminated at aphelion, and therefore receive less heat.

Temperatures on the planet thus vary enormously. At perihelion, when Mercury is approximately 4.8 million km from the Sun, the surface temperature at the subsolar point soars to about 427°C or 800°F, high enough to melt zinc. Around dawn it plummets to a stark -183°C, roughly 300° below zero on the Fahrenheit scale. The reasons for this steep temperature gradient, which is further proof that Mercury has similar surface properties to the Moon, are the length of the planet's day and the intense solar radiation.

Another important consideration is the absence of a significant atmosphere. This almost certainly disappeared at an early stage in the planet's history, as a result of high surface temperatures and the planet's low escape velocity. Traces of sodium were detected by Earth-based spectroscopy in 1985, but the amount is subject to fluctuation because of interaction with ultra-violet radiation from the Sun. Relatively high levels of helium, perhaps arising from radioactive decay or capture from the solar wind, were also registered by Mariner 10.

With a diameter of 4878 km Mercury is intermediate in size between the Moon and Mars. Yet its mean density is 98% that of the Earth. This is remarkably high, and implies that Mercury has a disproportionate share of heavy elements, possibly because it formed

*2. Eighteen separate images were used to make this picture of Mercury. The images were
made by Mariner 10 on 31 March, 1974.*

close to the Sun. Astronomers believe it is composed of about 70% metals and 30% silicates. Since iron is the most cosmically abundant heavy element, it is assumed that the planet has a partly molten iron core, roughly equivalent to the Moon in diameter. This is surrounded by a relatively thin shell of silicate mantle and crust, about 600 km thick.

There is no direct seismic data relative to the internal structure of Mercury, but a large metallic core is necessary to explain the significant, though weak, magnetic field detected by Mariner 10. Even so, the core is unusually large. The conclusion is inescapable. Both the chemistry and structure of the mantle and lithosphere may be significantly different from that of the Earth. This has implications for surface relief and the processes that formed and modified it.

Mariner 10 imaged about 45% of the surface down to a resolution of 1 km during its three passes of the planet in 1974 and 1975, to reveal the record of intense late bombardment already suggested by ground-based photometry and other studies, besides similarities to certain morphological units on the Moon. However, differences were apparent, notably in the spread of ejecta deposits on Mercury, attributed to its higher surface gravity.

As on the Earth, Moon and Mars, distribution of the major physiographic provinces is asymmetric. One side is heavily cratered and resembles the lunar highlands. The opposite side, photographed during the outward bound phase of the flyby, shows both cratered relief and extensive smooth plains reminiscent of the lunar maria. A number of randomly

3. *The Caloris Basin, the largest impact basin on Mercury's surface, is on the right of this image. The ring of mountains around the edge of the basin indicates that the centre of the basin is just off the edge of the picture.*

sited, multi-ringed impact basins were also observed. The 2000-km diameter Caloris Basin is the largest of these. It is a magnificent mountain ring, located in the northern hemisphere near one of the 'hot poles', and has peaks rising 2 km above the mean surface level.

Evidence of tectonic activity with possible implications for global contraction exists in the form of scarps. These are peculiar to Mercury, and show as long, sinuous cliffs, or low breaks in the topography, which extend for hundreds of kilometres, and rise to heights of from a few hundred metres up to 3 km. One explanation is that they formed early, as Mercury cooled and contracted, causing the thin, brittle crust to fold and crinkle.

Prior to Mariner 10, Earth-based optical and radar investigations had determined the density, mass, radius and rate of axial spin of Mercury. Its surface was also understood to resemble that of the Moon. In the aftermath of Mariner 10, the image is sharper. Mercury is still elusive, but no longer enigmatic. It is a curious, shrunken little world of unexpected contrasts; Earthlike on the inside, Moonlike on the outside. A world of harsh extremes whose surface is starkly inscribed with the signature of a violent past.

Venus

Venus, the familiar morning and evening star, blazes out of the twilight to outshine Sirius and rival Jupiter. Of all the larger planets it is the one most similar to Earth, and on occasion is the third brightest object in the heavens. Like Mercury it oscillates backward and forward, appearing first to the east and then to west of the Sun, but never straying more than 47° from that body, and sometimes remains visible up to $4\frac{1}{2}$ hours after sunset. Venus completes a circuit of the Sun once every 224.7 days at a distance of between 0.72 AU and 0.73 AU. The least Earth–Venus distance is about 44 million km and occurs at inferior conjunction. Venus then subtends an apparent angular diameter of 64 arcsecs, but disadvantages direct visual observation by having its night side towards the Earth.

With a diameter of 12 104 km, and a density comparable to that of Earth, Venus is statistically the Earth's twin. This fuelled the imagination of the early telescopists who hoped for startling appearances. Venus showed nothing of the sort. Galileo discovered the phases in 1610, and helped to establish the Copernican theory. The presence of an atmosphere was deduced by Mikhail Lomonosov in 1761, although it had been suspected by Christiaan Huygens over six decades earlier. But all attempts to determine the planet's rate of axial spin ended in confusion and acrimony. Even the enigmatic satellite, so often reported in the seventeenth and eighteenth centuries, was ultimately declared apocryphal. Nothing could be affirmed. Speculation flourished, and Venus successively became a swampworld; a windswept desert; a Seltzer sea; and a petroleum ocean.

By the early 1960s radar and robotic scouts had transformed the uncertain vision of the telescope. Venus was more like the classical concept of Hell – astonishingly hot, swathed in a dense veil of corrosive gases, with a surface that glowed from its own heat. Even that mystery of mysteries, the rate of axial spin, was finally resolved when radar beams bounced off the surface of the planet to reveal a 243-day retrograde rotation.

All we ever see of Venus through an optical telescope is the upper deck of an unbroken sheet of lemon-yellow cloud, which reflects about 79% of incident light; the highest known albedo in the local planetary system. Featureless in integrated light, this surface shows dusky blotches and bright bands in the ultra-violet. First imaged by Frank Ross in 1927, they are characterised by a windswept look. The most obvious is like a letter 'Y' laid sideways. However, the significance of these markings was not immediately appreciated. In 1957 the French amateur astronomer Charles Boyer re-examined the planet in the ultra-violet. He verified the earlier findings, and discovered the markings have a four-day rotation period. It was a major breakthrough. For whatever their cause, the ultra-violet markings provide important clues about Venus's weather machine. Their rapid motion, in particular, which is estimated at 100 m/s at the cloud tops in the equatorial regions,

4. *Venus imaged by the Pioneer Venus Orbiter on 14 May, 1988. The picture shows the top layer of clouds surrounding Venus. Sunlight is reflected from the cloud tops, the lighter colour indicates the brightest reflection.*

indicates the presence of global-scale winds possibly driven by a diurnal cycle of insolation much longer than Earth's.

Venus's thick, hot, carbon-dioxide atmosphere is huge. It extends up to 250 km above the surface, but roughly 90% of its total volume lies below 28 km. Surface pressure is almost one hundred times that on Earth, and the temperature at the surface is around 480°C or 890°F. Below the clouds conditions are uniform and oppressively hot, quite impervious to even small changes in solar insolation. Thus the climate of Venus is probably very stable.

Space-craft data indicate three distinct cloud decks underneath a variable haze layer, in the altitude range 45 to 65 km. Earth-based polarimetry and spectroscopy identify concentrated sulphuric acid droplets and sulphur crystals as the main constituents of the uppermost deck. At 65 km it is thought to be the source of the ultra-violet markings.

Of greater significance is the observed migration of the ultra-violet markings towards the poles. This has the characteristics of a Hadley cell, the simplest circulation system to exist in a planetary atmosphere. Hot air rises at the equator, cools, and is transported towards the poles, where it descends to be recycled at a lower level by a pole-to-equator return flow. Dramatic evidence of Hadley circulation in the form of a polar cloud clearance was provided by Pioneer Venus Orbiter (P V O) infra-red measurements.

Another feature detected by infra-red measurement is the circumpolar collar. This is a current of cold air enveloping the pole at a radial distance of about 2500 km. It is 1000 km across and 10 km thick in the vertical direction. Temperatures inside the collar, where the cloud is thinner, are markedly lower than at equivalent external heights.

Soviet Venera space-craft gave us our first direct sight of the planet's surface. They revealed a stony, desert landscape: smooth, flat-topped outcrops and large angular flat slabs, showing dark patches suggestive of chemical erosion. Composition of venusian soil resembles that of basalt found on terrestrial sea-beds. Large circular features, interpreted as craters, were first detected by Earth-based radar facilities at Arecibo, Puerto Rico, and Goldstone, California. The most comprehensive data about surface roughness, however, has come from the P V O mission, which carried a radar altimeter and gave coverage at a horizontal resolution of about 100 km, and a vertical resolution greater than 200 m. This data set gives us our first global picture of what lies beneath the cloud.

Venus emerges as uniformly flat with comparatively few eruptions of highland. Physiography can be subdivided into smooth, rolling plains (70%), lowlands or depressed areas (20%), and highlands (10%). Rolling plains include a diversity of features, such as volcanoes, craters and rifts. Lowlands are relatively smooth, basin-like units, of which the most conspicuous is Atlanta Planitia, located in the northern hemisphere. In size it is equivalent to the Gulf of Mexico, and is roughly 1.4 km below mean surface level.

Ishtar Terra, Aphrodite Terra, Beta Regio and Alpha Regio are the principal highland units. Ishtar Terra is unlike anything else discovered on the terrestrial planets. Located at latitude 70°N, it rises steeply to a height of several kilometres above the plain, and is about equal to Australia in extent. On its western part is a vast plateau (Lakshmi Planum), 2500 km in diameter and 3 to 4 km high, encircled and cut by tall mountain ranges up to 3 km in height.

Stretching 1000 km along and south of the equator stands Aphrodite Terra. About

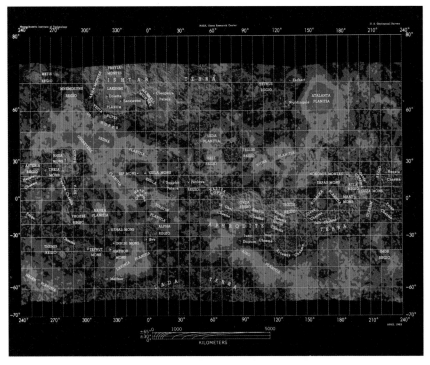

5. *This contour map covers about 83% of Venus's surface. It was produced from data from the Pioneer Venus Orbiter. Before Pioneer less than 1% of Venus's topography had been measured. Highland areas are like continents themselves. Ishtar Terra in the north is the size of Australia.*

equal in size to Africa, it is even more rugged than Ishtar. It is made up of mountains to the east and west, separated by a lower region, and includes an enormous circular feature, 2400 km in diameter, possibly a volcanic caldera. Still more dramatic are the gigantic trenches or rifts. The most remarkable is Diana Chasma, a huge rift valley over 1000 km in length. It has a depth of 2 km below the mean surface, and a width of about 300 km, and may well be as remarkable as the Valles Marineris on Mars.

Venus's thermal evolution and the nature of its interior are poorly understood. The planet evidently has a crust, mantle and core, but there are no positive data. However, volcanic features of recent origin were imaged by Magellan, the latest US mission to the planet, and ongoing vulcanism is indicated. In contrast to Earth, Venus does not have a magnetic field of internal origin. A weak field induced in the ionosphere is probably caused by interaction with the solar wind.

The Earth and Moon

Earth is the largest and densest of the inner planets, remarkable for its ubiquitous water and the presence of living organisms. It has an equatorial diameter of 12 756 km and is

6. *Earth from space, imaged by Apollo 17. Africa is clearly seen in the upper part of the globe.*

the third planet in order of distance from the Sun, around which it circles once every 365.25 days at a mean distance of 149 597 870 km, spinning on its axis in a period of 23 hours, 56 minutes and 4 seconds. Viewed from space it looks cool and inviting: a world of swirling white clouds and polar snows, blue oceans, and brown and green land masses, housed in an atmosphere rich in nitrogen and oxygen. All of which signifies a complex surface environment and a sphere of interaction difficult to characterise by a simple set of equations.

Water permeates the solid crust to a depth of several kilometres, and covers over two-thirds of the surface in vast oceans. Large amounts are also locked into the polar ice caps and in the atmosphere where it appears as haze and cloud. By acting as a heat reservoir, and promoting the secondary circulation of heat, this great body of water (hydrosphere) is important in ameliorating the weather and climate of the planet.

Water is also essential to much of the Earth's abundant variety of life (biosphere). The living organisms themselves range from the simplest cell to highly evolved intelligent animals, and are found in locations as diverse as the polar caps and ocean floors. Photosynthesising plant organisms invest large parts of the surface, and provide much of the molecular oxygen that comprises about 20% of the atmosphere. The ozone layer, which screens the fragile biosphere from the harmful effects of the Sun's ultra-violet radiation, is derived from this oxygen, and is concentrated between 12 and 50 km above the surface.

Stripped of its covering of air and water, the Earth's topographic surface is observed to consist of two major provinces: ocean basins and high-standing continents. The

7. The crescent Earth is seen rising from behind the Moon in the foreground.

transition between the two regimes is relatively sharp. Much of the elevation is either near sea level or about 4 km below. Regions of markedly higher or lower profile are limited in extent; the highest units are mountain chains, the lowest, oceanic trenches. Oceanic crust is about 6 km thick, relatively young and made up of high density igneous rock of basaltic composition. This suggests rapid recycling and the almost complete destruction of all record of the older crust. In contrast, continental crust (35 km thick), composed of igneous, metamorphic and sedimentary rocks, is older and buoyant. Thus, showcased in Earth's topography is a global pattern of internal and surface activity, which, if our interpretation of the post-late bombardment fossil record is correct, indicates the continents have grown by addition from underlying resources, accretion of island arc volcanic systems, and interaction with ocean basins.

The dynamics involved are described by plate tectonics theory. Briefly this envisages Earth's apparently rigid outer layer, the lithosphere (100 km thick), to comprise a number of segments or plates that move laterally relative to each other over a mobile layer called the asthenosphere. Plates are created at mid-ocean ridges by upwelling magma that divides and flows outwards by a process known as sea-floor spreading. Ridges thus represent divergent plate boundaries. Where plates converge, the boundary is typified by a deep oceanic trench, an associated island arc, and intense seismic activity, as one plate plunges or subducts beneath the other into the mantle below. Over millions of years the continents change their relative positions as they are carried along by the plates in which they are embedded. When they collide, or meet oceanic crust, mountain ranges tend to form. Thus, plate tectonics is one of the major processes actively shaping and changing the face of our planet, and provides an explanation for some of the most dramatic and violent events in geology.

Earth's internal structure consists of a series of concentric layers distinguished by compositional differences. The dense nickel–iron core, approximately the diameter of Mars, has a molten outer zone and a rigid nucleus, and is surrounded by a thick mantle of iron, magnesium and silicates. The thin, outermost layer (lithosphere), is about 100 km thick and is composed of silicates and dense basaltic rocks.

Convective currents in the molten outer core may be related to the generation of the planet's magnetic field, which takes the form of a magnetic dipole. This interacts with the solar wind and forms the Van Allen Radiation Belts, which completely encircle the planet.

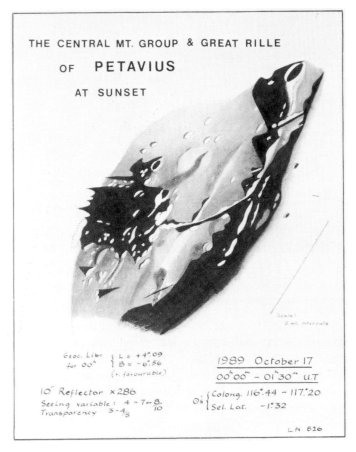

8. *A detail of the Moon's surface drawn by Harold Hill from observations made on 17 October, 1989.*

9. *The Moon showing maria, highlands and craters, photographed by Ron Arbour. The craters are best seen near the terminator. Mare Tranquillitatis is the dark maria just above the centre of the image, Mare Serenitatis is above and to the left of it.*

Their existence was unsuspected until the late 1950s and their discovery was the first important finding to emerge from the exploration of interplanetary space.

Although the atmosphere extends for over 500 km above sea level, 75% is restricted to its lowest layer, the troposphere. This extends from sea level, where the pressure is 1013 mbar, to the tropopause at a height of 8 km at the poles and 18 km at the equator. Cloud and weather systems occupy its lowest levels.

Earth's weather machine is triggered by incoming solar radiation. Temperature differences produce pressure gradients, and, once these are established, winds are generated and atmospheric circulation is set in motion. The actual process is very complex, and is ultimately a function of surface relief and interaction between the oceans and atmosphere. There is fossil evidence of climatic change over great intervals of time. Various causes are proposed, including continental drift, changes in the chemical composition of the atmosphere, and variations in the eccentricity of the Earth's orbit.

The Moon is Earth's sole natural satellite. One of the largest satellites in the solar system in comparison to the size of its planet, the Moon probably formed by accretion starting around 4600 million years ago, but has been inactive for the past 2000 million years. It is less dense than the Earth, has no atmosphere and lacks a hydrosphere. Since it has synchronous rotation, only about half the surface is visible from Earth. The lunar surface is broadly divisible into two types of region: the younger, low-lying dusky maria, and the ancient terrain of the bright, heavily cratered highlands. Because it has largely completed its evolution, and is exceptionally deficient in volatiles, the Moon provides clues to the evolution of planets in general, and is therefore a focus for planetological studies.

Our quest to understand the Earth and its satellite, its present and its past is broad-based. First, by studying our immediate environment we can better appreciate what has happened or might be happening on other worlds. Secondly, without knowledge of the past and the present we cannot, with confidence, predict how our actions will affect the future. For Earth is a sensitive organism; adversely affect just one part, and the ripples spread to cause unpredictable consequences.

Mars

For centuries astronomers puzzled over the small orange-red ball that glowed in their eyepieces; so familiar were they with its polar snows and seasonal changes, its clouds and mists, that in 1793 William Herschel exclaimed: 'The analogy between Mars and the Earth is perhaps the greatest in the whole solar system.'

Its brightness, its baleful hue, the extent and eccentricity of its motion among the stars, all combine to make Mars one of the most remarkable, indeed compelling objects in the heavens. Aristotle watched its occultation by the Moon, 4 May, 357 B C, and inferred its greater distance. Johannes Kepler using observations bequeathed to him by Tycho Brahe, established its true orbit, and deduced his celebrated laws of motion.

Mars moves around the Sun in an ellipse, the semi-major axis of which is 1.52 times that of the Earth's orbit. The eccentricity of the orbit is 0.093; hence the actual distance of Mars from the Sun varies between 1.381 A U and 1.666 A U. Mars takes 1.861 years to complete its orbit. Its synodic period is about 760 days, but because of the orbital eccentricity and the fact that successive oppositions occur in different parts of the planet's orbit, the distance between Earth and Mars varies widely. At perihelic oppositions it can be as small as 55 million kilometres; at aphelion as great as 100 million kilometres. Accordingly the apparent diameter of Mars viewed from Earth varies inversely as the distance. At a perihelic opposition the disc appears almost twice as large as at an aphelic opposition.

Midway in size between the Earth and the Moon, Mars measures 6794 kilometres at its equator, and is 35 kilometres less in the polar direction. Its day is 37 minutes longer

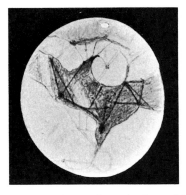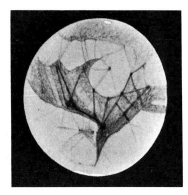

10. Two views of Mars drawn by Percival Lowell in his observing book. They clearly show his 'canals'.

than Earth's, and its axis is tilted 25° to the plane of the planet's orbit. At perihelion the southern hemisphere is tipped towards the Sun; southern summers are thus hotter and shorter than those in the opposite hemisphere.

Mars gave new meaning to the term 'extraterrestrial life' in 1877 when Schiaparelli discovered its surface was covered by a network of fine shadowy streaks. Science hesitated, pondering a momentous possibility. In the meantime the media headlined the news and the Martians insinuated themselves into our culture. Ultimately they became the authentication, and the motive of sustained interest in the planet.

A desolate, crater-scarred analogue of the Moon thus took scientists by surprise in July 1965 as Mariner 4 relayed the first close-up pictures of the planet back to Earth. Was the reaction justified? Not really. Craters had been predicted in 1951 by Clyde Tombaugh and Ernst Öpik. They had even been spotted in 1917 from the Yerkes Observatory by the American telescope-maker and comet-discoverer John Mellish. Earlier still, in 1894, E. E. Barnard had correctly inferred the true character of the martian surface relief.

Mariners 6 and 7 arrived at Mars in 1969 to extend the coverage of Mariner 4. Medium-resolution images of the albedo detail were returned, along with relief imagery taken at close range. Two crater types were quickly recognised, but the etched and pitted polar regions, and chaotic terrain near the equator, all pointed to one conclusion: the impressions from Mariner 4 were quite misleading. Mars, in fact, was neither Moonlike, nor yet did it parallel the Earth.

Knowledge jumped dramatically with the arrival of Mariner 9 at the planet on 14 November, 1971. Mars was at perihelic opposition and beset by a ferocious global dust storm. Geologists waited gloomily. Atmospheric experts, on the other hand, utilised the changing patterns in the dust clouds to improve their data about martian winds. As the storm eased, the waiting observers watched in astonishment as the Tharsis volcanoes broke through the dusty air. The largest, subsequently named Olympus Mons, corresponded to a dusky spot previously charted as Nix Olympica. Excitement rose to fever pitch with the discovery of a giant canyon complex, even more spectacular than the Grand Canyon. Valles Marineris, as it is now called, stretches more than 4000 km along the martian

11. *Mars from Viking 1 on 17 June, 1976. The large volcano, Olympus Mons, is toward the top of the picture. Close by is a row of three other volcanoes.*

12. *Two Viking space-craft landed on the surface of Mars in 1976. This is the first view that man had of the martian surface.*

equator. It features as a canal on the old albedo charts, where it was named Coprates. Both the volcanoes and the canyon complex are of more recent origin than the craters.

The Viking mission reached Mars in 1976; its purpose to answer the many questions raised by the Mariner series. Orbiters were programmed to make a comprehensive survey of the whole planet. Two landers were set down on the surface, one in Chryse Planitia, the other in Utopia Planitia. These were to determine the chemistry and composition of the crust, detect seismic activity, monitor the weather, analyse the atmosphere, and conduct biological experiments in an attempt to search for evidence of indigenous life forms. Viking marked the close of the first martian reconnaissance.

As on the Moon, Earth and Mercury, the distribution of the major physiographic provinces is asymmetric along a line inclined about 35° to the equator. Smooth, thinly cratered volcanic plains characterise the northern hemisphere. Amidst these barren, rocky deserts Viking Lander 1 scanned dust dunes, rolling hills, exposed bedrock, and pumice-like rocks recumbent beneath a salmon-pink sky. Northward, on fractured ground, in the ejecta field of a young crater, Lander 2 looked out over a flat, rocky expanse devoid of dunes, hills and rocky outcrops.

Heavily cratered and generally at a higher elevation than its counterpart, the southern hemisphere closely resembles the lunar highlands, and includes large circular basins, of which 1600 km diameter Hellas is the most conspicuous. It is partially eroded, and at 6 km deep is probably the lowest part of the planet.

Possibly the most enlightening feature on Mars is the Tharsis bulge. It swells up between the hemispheres to embrace about one-quarter of the surface, and is surmounted by huge shield volcanoes. Of these, Olympus Mons is the highest, standing 27 km above the mean surface on a base 600 km across. It features as a dark spot on the old albedo maps, where it is called Nix Olympica. Three more such structures, Arsia, Ascraeus and Pavonis, straddle the summit line of Tharsis. Each is capped by a caldera several hundred kilometres in diameter. Alba Patera, northeast of Olympus Mons, is the second largest volcano on Mars. It is a huge but inconspicuous lava outflow, some 4 km high, that extends for about 1800 km. Cracks and radial fault systems about Tharsis testify to the enormity of the geological forces that led to its formation.

Mars has a tenuous carbon dioxide atmosphere one-hundredth the density of Earth's, and is known to experience episodes of high-velocity winds. These are capable of raising planet-wide dust storms, and are the principal mechanism of erosion and deposition on the planet. Although martian meteorology is relatively uncomplicated by oceans and high mountain chains, its unique characteristics cannot be neglected. Topography varies in altitude by about 30 km. Pressure at the summit of Olympus Mons, therefore, will be an order of magnitude less than at its base.

Although water vapour exists at the polar caps and is an essential component of the martian atmosphere, free water is unknown on Mars. Remarkable evidence of possible glaciation, and ancient flooding on a large and massive scale, in the form of a network of meandering valleys and tributaries, which look suspiciously like flow channels, was discovered in the Chryse plains by Mariner 9. What happened to the water on Mars, however, is one of the major puzzles to confront planetologists. Of course significant amounts of volatiles may be trapped below the surface. The episodic, or temporary, nature

13. *This view is of the larger of Mars' two moons, Phobos. It is about 27 km from end to end, and like its companion Deimos it travels in an almost circular orbit close to Mars' equator.*

of fluvial activity on Mars, as well as the evidence for glaciation, raise basic questions about climatic change on the terrestrial planets, including the Earth.

The martian polar caps are of two kinds, seasonal and permanent. The latter are composed of water ice, and are of long duration. The waxing and waning of the former, however, indicates that the atmosphere cycles carbon dioxide between high and low latitudes on a seasonal basis. The northern permanent cap covers much of the region above 80°. The less extensive southern field is centred on 87° latitude, and is offset from the geometric pole.

Equatorial temperatures range from 26°C soon after noon in summer to −111°C prior to sunrise. Yet, despite these low temperatures and the paucity of water vapour, the martian air is comparatively moist. Clouds and mist are common in the salmon-pink sky. During the spring and summer orographic cloud is associated with the Tharsis volcanoes. There is also low level haze, typically at dusk and dawn, together with fogs that linger in the valley bottoms. A sharp edge to the planet is rarely seen, and thin high altitude hazes are common.

Viking found no evidence of organic life. With its ultra-violet flux and harsh arid

14. Preliminary observations made with the Hubble Space Telescope yielded the sharpest images of Mars ever taken from Earth or Earth's orbit. A thick canopy of bluish clouds covers the icy north polar regions, where it was martian 'winter' at the time of the observation.

condition, besides its tendency to oxidise, the martian environment is not one to encourage the development of organic compounds.

Mars has two small moons, Deimos and Phobos, both discovered by Asaph Hall in 1877. Deimos is the outermost and is locked in synchronous rotation with Mars. It orbits in a period of 1.262 days at a distance of 23 500 km. It is irregular in shape (15 × 12 × 11 km) and has a 7% albedo. Phobos, the larger of the two, measures 27 × 21 × 19 km. It orbits at a distance of 9380 km in 0.319 days. Viewed from the surface of Mars, it rises in the west, speeds across the sky at a rate of 47° per hour, and 5.5 hours later sets in the east. Phobos also has a low albedo (5%), and is dark grey in colour.

If Mars in its gross appearance resembles the Earth, it is now clear that the analogy as originally envisaged is nullified. Its evolutionary history is distinctly different as demonstrated in its tectonics. Whereas the Earth is dominated by plate movements, on Mars the crust is fixed and evidences no sign of those features which on Earth indicate interaction between the plates.

The age-old dream has become a reality. We have triumphed over gravity. From the great central sweep to the outermost reaches of our system, the alchemy of new technology has transformed the uncertain tradition of the telescope. We have touched the inner planets; surveyed, mapped, sampled and measured in a progress of unequalled discovery. They are no longer specks of light wandering through the stars.

With Mercury the questions are still broad, first order and exploratory. Venus cloaks its horror in light, but its significance is inescapable. Mars, on the other hand signifies yet another scenario, but whether of Lowellian import we cannot judge. Either way the questions posed are critical to the understanding of the formation of the Earth, its atmosphere and oceans, and the physical and chemical foundations of life itself. Comparative planetology is not just an academic exercise. It is of critical interest to everyone since it provides telling insights into the very future of our home planet. It is moreover a venture in which amateur and professional can work together to mutual benefit.

3

Jupiter

JOHN ROGERS

Images of Jupiter reveal a world totally different from the planets nearer the Sun. It is a vast flattened globe of gas, on which the only features that we can see are multi-coloured clouds, drawn into parallel bands by the planet's rapid rotation. As the largest planet, with a mass 318 times that of the Earth, Jupiter has kept all the gases and some of the heat that it had when it formed. In fact, it is the largest object that we can study short of the stars themselves. It may also be a representative of a very abundant class of objects in the universe, since the most conventional candidates for supplying the 'missing mass' of the galaxies would be planet-sized objects.

Jupiter's equatorial diameter is 142 800 km, eleven times that of Earth, but its rotation period is so short (9 hours 55.5 minutes) that the poles are noticeably flattened, with a polar diameter of 134 200 km. The thick and cloudy atmosphere is very dynamic, with a complexity of structure and motion unmatched on any other planet. Because of the rapid rotation, winds are channelled eastward and westward along lines of latitude, dividing the atmosphere into dark bands called belts and bright bands called zones. This structure sets the scene for an ever-changing display of disturbances in the clouds, changing over years or weeks or even days.

The main belts and zones can be seen in a small telescope, and a slightly larger telescope, of 8 cm aperture, will also show some of the irregularities and spots. With telescopes of 15–35 cm, amateurs can see many of these and study how they change. Spots have been

1. *Jupiter drawn by E. M. Antoniadi on 8 December, 1928. Just above the equator, the chaos of spots comprises a South Equatorial Belt Revival. The black spot on the left is a satellite shadow.*

seen on Jupiter since 1664, and the difference between equatorial and higher-latitude currents was discovered in the seventeenth century. More detailed observations began in the nineteenth century, particularly around 1878 when astronomers' attention was attracted by the extraordinary redness of the Great Red Spot. Since that time, amateurs have maintained a continuous watch on the planet. Amateur observations, especially by members of the British Astronomical Association, had defined many of the different currents and types of disturbance by the end of the 1920s. Professional photographic surveys were mounted in the 1960s and 1970s. Visual observations continue to reveal further dramatic and intricate patterns in the behaviour of the atmosphere.

Now our view of the planet has been transformed by space-craft: Pioneers 10 and 11 in 1973 and 1974, and Voyagers 1 and 2 in 1979. Most of the space-craft experiments were designed to study the magnetic field, charged particles and radiation round the planet, which previously were only known from the planet's radio emission. But they also contained instruments for studying the infra-red and ultra-violet spectra and the visible polarisation of the clouds, so as to measure their heights and temperatures, and of course they also carried cameras. The Pioneers' 'cameras' were comparatively crude scanning photometers and only surpassed Earth-based resolution on the day of encounter, but the Voyagers' 'cameras' were rather powerful telescopes, producing spectacular pictures over several months. These pictures were compiled into time-lapse films which dramatically revealed the patterns of motions in the atmosphere, as well as showing internal structures of disturbances which Earth-based observers had tracked for many years. The next probe to Jupiter, launched from the US Space Shuttle in October 1989 and due to arrive in December 1995, is Galileo. One part of it will orbit the planet, studying Jupiter and its moons at close quarters for a year or more; the other part will be the first probe into Jupiter's atmosphere, sending back direct measurements in and below the visible clouds as it descends towards its destruction.

The structure of the atmosphere

The visible surface of Jupiter's clouds has been studied in detail. The motions of the clouds have been tracked by amateurs, and their physical nature has been deduced by professional astronomers and chemists from observations at non-visible wavelengths. But what happens under the clouds is still a matter for theory and speculation.

The planet's atmosphere – indeed, the whole planet – consists mainly of hydrogen and helium. The atmosphere also contains substantial amounts of ammonia and methane, as well as water, simple hydrocarbons, carbon monoxide, phosphine, and other gases.

The nature of the clouds has been deduced from spectroscopic and occultation measurements, which indicate the temperature, pressure and chemistry of the atmosphere at different levels. The famous orange clouds of the Great Red Spot and some other regions are the highest; it is still not known what produces this colour. The next highest clouds are those of the white zones and white spots; the whiteness of these clouds, and the temperature and pressure at which they occur, imply that they must be made of ammonia ice crystals. Brown belts and spots are lower; these clouds are probably made mainly of ammonium hydrosulphide. The deepest visible levels are the dark bluish-grey patches near the equator. At these levels, which are the warmest as well as the deepest observed, there may be extensive water clouds not unlike those on Earth. Of course, the other gases present are utterly unlike those we breathe, and could not support life as we know it; but the jovian gases may be able to undergo chemical reactions analogous, in some ways, to those which produced life on Earth.

Below the clouds, there is no solid surface. An imaginary (and indestructible) observer descending into the planet would pass through denser and denser gases, pressed by the planet's tremendous gravity and gradually passing into the liquid phase, until the pressure was around three million times that at the Earth's surface. At that point, perhaps 15 000 km below the visible surface, it is believed that the hydrogen becomes a liquid metal, forming a thick mantle in which the planet's powerful magnetic field arises. At the centre there is a dense core, comprising a few per cent of the planet's mass, and probably made of rocky elements like those of the Earth.

The planet is hot inside; it emits twice as much heat as it receives from the Sun. It must have been hotter when it formed, thanks to the energy released as it collapsed from a gas disc, and it has been cooling down ever since. This heat must be the main driving force for the powerful motions seen in the atmosphere.

What is the pattern of these motions? Clues to the dynamics of the atmosphere have been coming together for many years, but the overall picture is not yet complete. As the winds blow mainly in currents along lines of latitude, amateurs have been able to measure them by repeatedly measuring longitudes of spots. This is simply done by recording the time at which a spot crosses the central meridian of the disc, as it is carried round by the planet's rotation.

Measurements of this sort revealed that many latitudes have characteristic currents of up to 1° longitude per day ($\simeq 10$ m/s), and that the whole equatorial region moves eastward in a broad, powerful current at 8° per day ($\simeq 100$ m/s). More rarely – sometimes at intervals of many years – distinct outbreaks of small dark spots on the edges of certain belts reveal much narrower rapid currents, or jetstreams, moving eastward or westward at several degrees per day (tens of metres per second).

The spectacular time-lapse films from the Voyager space-craft showed that these jetstreams run continuously, and that there is a regular pattern of jetstreams over most of the planet. Each belt has an eastward jetstream on the side towards the equator, and a westward jetstream on the side towards the pole. These patterns fit in neatly with the altitudes of the belts and the laws of motion on a rotating body. The belts must be low-pressure areas, with an overall sinking of gases, bounded by jetstreams that constitute a cyclonic circulation. The zones must be high-pressure areas, capped with high-altitude white clouds, bounded by jetstreams with an anticyclonic sense.

The same principles apply to individual spots. Professional Earth-based photography in the 1960s revealed that the Great Red Spot has an anticyclonic circulation (anticlockwise, with a period of 6–12 days), which fits with the fact that it is a high-altitude cloud system overlying a zone. The Voyager films showed that almost all the smaller spots also had rapid circulations, anticyclonic in the zones and cyclonic in the belts. Thus the spots roll between the jetstreams like ball-bearings.

What we do not know is how deep these motions go. A wide variety of theoretical models can produce circulations more or less resembling those on Jupiter. In some of these models, the motions are almost confined to the visible cloud layers; in others, they extend down all the way to the edge of the metallic hydrogen mantle. At present there is no way of choosing which model is right.

Large-scale disturbances in the atmosphere

Even though the origin of the disturbances in Jupiter's atmosphere is not fully clear, the patterns which they follow can be described in detail, thanks to the synthesis of data from amateurs and from space-craft.

There are several characteristic types of feature, generally called 'spots'. Some are sharply defined bright ovals, sometimes with a very dark border; these are high-altitude anticyclonic clouds. Other spots are elongated dark bars, dark grey or reddish-brown, which are cyclonic like the belts in which they are embedded. Some belts also contain light-coloured cyclonic formations, which are often convoluted and turbulent. In the equatorial region are different types of features: dark blue-grey patches, as described above, adjacent to brilliant white plumes which stream out towards the equator.

The most famous spot is the Great Red Spot (Figures 2 and 3), which has been observed certainly since 1831 and possibly since 1665. In some years it is a dark orange oval; at

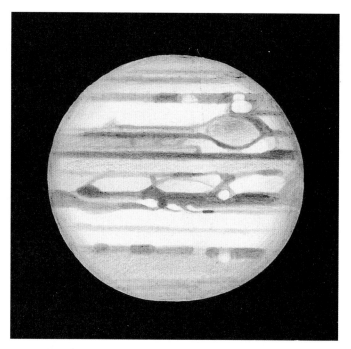

2. *Jupiter drawn by John Rogers on 13 Feburary, 1979, showing the Great Red Spot and an ochre coloration of the Equatorial Zone.*

other times there is no 'spot' as such, but its oval form is still evident as an indentation in the adjacent belt. Visual observers have followed the Spot's changes from an intense brick-red to a pale grey and back again, and its slow movements back and forth around the planet; and they have shown that these changes are often caused by events in neighbouring regions. Space-craft have accurately measured the height and speed of the Red Spot's clouds, and theoretical physicists have worked out that a great anticyclone like this can sustain itself indefinitely on Jupiter.

There is only one Great Red Spot, on the South Equatorial Belt (SEB), and it influences motions around almost half the planet. But on the next belt to the south, the South Temperate Belt (STB), there are three similar oval spots. They are smaller, and white rather than red, but they have identical circulation patterns according to space-craft pictures. (See Figure 3). Amateurs watched them form around 1940: the bright zone south of the belt became divided into three sectors, which gradually contracted to form the three white ovals, nestling in the south edge of the STB. Their contraction has continued for fifty years and they may soon disappear.

On the next belt to the south again, there are more white oval spots of the same type, but they are again smaller and more numerous than those on the STB.

A 'South Tropical Disturbance' is a related type of disturbance, in the same latitude as the Great Red Spot, and the greatest on record existed from 1901 to 1939. It involved a complete reconnection of jetstreams on opposite sides of the zone. To the astonishment

3. *Voyager 1 photograph of the southern hemisphere of Jupiter, including the Great Red Spot with one of the South Temperate white ovals above it. Taken on 3 March, 1979 (18 days after Figure 2)*

of observers at the time, this dark shading across the zone blocked the passage of spots moving with the jetstream alongside the SEB, and diverted them onto the opposite jetstream alongside the STB, so that they returned the way they had come! Fortuitously, a new South Tropical Disturbance developed in 1979 just as the Voyager space-craft were scrutinising the planet, so they photographed it in great detail (Figure 4). This one was created out of a complicated interaction between the Great Red Spot, a STB white oval, and a cyclonic structure in the STB. It was watched from Earth for another two years until it disappeared. A similar interaction, six years later, produced the 'little red spot' shown in Figure 5.

Sometimes a whole belt or zone undergoes a change all round the planet. A belt may disappear, becoming covered with white clouds. For example, at the time of the Pioneer space-craft encounters, most of the SEB had disappeared. By the time of the Voyager encounters, the SEB had revived as described below, but a sector of the STB had become whitened out. Alternatively a zone may adopt a strong ochre colour for several months

4. *Voyager 2 close-ups of a major disturbance (a South Tropical Disturbance) on Jupiter in June 1979. Small vortices become caught up in the larger vortex, whose anticlockwise circulation can be seen in the ten hours between these two photographs.*

5. *Jupiter on 10 July, 1986, photographed from the Pic du Midi by Jean Dragesco. This enhanced-colour composite was made by John Rogers. It shows the contrasting colours of the belts and dark bluish patches just below the equator.*

or years. This too had happened in the equatorial zone, which is the most common site of such a coloration, at the time of the Voyager encounters. The Voyager space-craft were extremely fortunate in the variety of activity which had developed on the planet at the time of their visits.

A quite different sort of disturbance affects the jetstreams along the edges of the belts. It consists of little dark spots that run along with the jetstream. In fact, the jetstreams are only visible from Earth when outbreaks of these spots are occurring. (It was spots of this type that revealed the circulation around South Tropical Disturbances.) The Voyager space-craft viewed such spots on two jetstreams and found that they were all anticyclonic eddies. Sometimes the spots form gradually along long stretches of belt, as if by instability in the rapid winds. On the other hand, sometimes they seem to be poured out from a single turbulent source. This occurs most spectacularly during a revival of the SEB.

SEB revivals are the most violent events known to occur on Jupiter. The revival is preceded by fading of the SEB, which becomes covered with white cloud, while the Great Red Spot stands out more conspicuously than ever. The revival typically begins with a dark streak and white spot across the latitudes of the SEB. From this source, turbulent dark and bright spots proliferate, moving in opposite directions on the jetstreams on opposite sides of the belt, and filling the middle of the belt with furious activity (see Figure 1). When the spots on the southern jetstream collide with the Great Red Spot, its dark red cloud deck is disrupted and it fades, while the dark 'hollow' around it reappears. Eventually the two branches of the disturbance overlap and the whole SEB is restored. These events have occurred at intervals of between three and thirty years during the twentieth century.

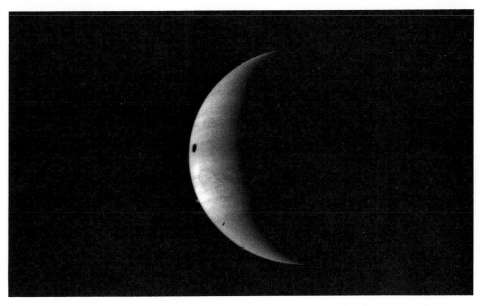

6. *View of the crescent Jupiter from Voyager 1. The large spot on the planet is a satellite shadow.*

The reader may have noticed that all the large-scale disturbances described so far have been in the southern hemisphere. The two hemispheres are indeed not symmetrical. The North Equatorial Belt does carry occasional 'little red spots', but they are much smaller and shorter-lived than their great archetype in the south; and the North Equatorial Belt also has had violent revivals, but not as well-organised as those of the SEB. Further north there are white ovals, but they also are smaller and more scattered than their southern counterparts.

The moons of Jupiter

Jupiter's four great satellites, the 'Galilean moons', were discovered by Galileo in 1610 when he first looked at Jupiter with a telescope. It was this miniature solar system around Jupiter which convinced him that the Earth was not at the centre of all planetary motions. For amateur astronomers today, these moons can be seen easily in binoculars, and they give a fascinating display of orbital motions. A small telescope will show them passing

7. *The four Galilean satellites; Voyager photographs to scale showing the relative sizes and colours. At top are Europa and Callisto, which is about the same size as the planet Mercury. Below are Io and Ganymede.*

in front of or behind the planet, casting their shadows on its clouds (Figure 1) or disappearing into the planet's shadow.

But Earth-based observers can resolve hardly anything on the surfaces of these moons. It was only during the epic twenty-seven hours of Voyager 1's cruise through the system, on 5–6 March 1979, that these moons were revealed as four new and varied worlds of ice and fire (Figure 7).

The outer two, Callisto and Ganymede, consist largely of ice around rocky cores. Their surfaces are ancient, peppered with the marks of impact craters, although many of the craters have almost subsided due to gradual creep of the ice. They also bear the marks of much larger impacts in the past, in the form of concentric rings looking like frozen ripples. Ganymede has also had internal activity, which has buckled large zones of its crust into wrinkled bands.

Closer to Jupiter is Europa, which is also made of water ice over rock, but its surface is almost completely smooth – except that it is criss-crossed by a network of long lines. At first sight these look just like the mythical canals that Percival Lowell believed covered the planet Mars, and Voyager 1 did not approach close enough to see more. But as seen at closer range from Voyager 2, more realistically, Europa looks like the Earth's Arctic Ocean. It probably has a mantle of liquid water, on top of which the ice crust cracks and shifts. Both the melting and the cracking of the ice are caused by tidal forces.

The most amazing moon is the innermost, Io. One of the most remarkable predictions in the history of science was published only three days before the Voyager 1 flyby, entitled 'Melting of Io by tidal dissipation'. In that article, S. J. Peale, P. Cassen and R. T. Reynolds predicted that 'consequences of a largely molten interior may be evident in pictures of Io's surface', and that 'widespread and recurrent surface volcanism would occur'. Their reasoning was that the orbits of Ganymede, Europa and Io are related in such a way that each repeatedly pulls on the others, with Io bearing the brunt of these gravitational tugs. The repeated flexing should heat up Io's interior and keep it partially molten. Voyager 1 revealed that this really happens.

So Io's surface as seen by Voyager 1 was not ice but a volcanic wasteland, with multi-coloured plains, sulphur lava flows, calderas, craggy mountains, 'frosts' of sulphur dioxide, and, above all this, the plumes of the continuously active volcanoes (Figure 8). The plumes are different from terrestrial eruptions and might be better described as huge sulphur-dioxide geysers, but in the low gravity and near vacuum of Io they spurt up for hundreds of kilometres and cover the moon with their deposits. And there are also eruptions like those of terrestrial volcanoes from time to time. Here is a world that surpasses the imaginings of science fiction.

Jupiter also has twelve much smaller satellites, in three groups of four. Two of these groups orbit very far from the planet, in highly inclined and unstable orbits, with mean distances of 11–12 million km and 21–24 million km; the orbits of the outer group are retrograde. Each group probably represents the fragments of an asteroid or pair of asteroids which may have been captured and fragmented in a single collision. The largest satellite in these two groups, Himalia, is only about 180 km across.

The third group of four orbits closer to the planet than Io. One member of this group, Amalthea, measures 270 by 150 km and was discovered telescopically. The other three

8. *Voyager 1 view of Io showing a volcanic plume on the limb.*

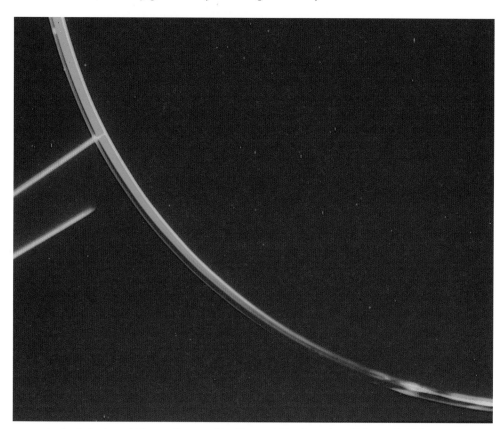

9. *Jupiter's ring which was discovered by Voyager 1 is shown here lit by sunlight from behind. The ring, at left, is partly blacked out by the planet's shadow. The larger bright arc is the twilit edge of the planet which shows false colours because the spacecraft was moving while the exposures were taken.*

are less than 110 km across, and are examples of the 'rocks' which the Voyager space-craft found in close orbits around all the giant planets. The innermost two orbits coincide with a tenuous ring of dust that was also discovered by the Voyagers. The ring extends from 129 000 km to 122 000 km from the centre of Jupiter, and is made largely of dust, which was lit up by the Sun from behind in the most striking space-craft photographs (Figure 9), although larger particles are also present.

Magnetic field and radiation belts

Like Earth, Jupiter has a strong magnetic field which forms an almost closed shell around the planet, deflecting the solar wind and trapping high-energy electrons and protons within it to form 'radiation belts'. These radiation belts were first detected in the 1950s by radio astronomers, who found that at long wavelengths Jupiter is one of the strongest radio sources in the sky. More recently they have been probed directly by space-craft. The radiation belts are much more intense than Earth's, so that a space-craft can only remain within them for a short time without suffering damage to its electronics. Pioneer 10 experienced one thousand times the lethal radiation dose for a human, and suffered some permanent radiation damage; Pioneer 11 and Voyager 1 both suffered from false commands induced by the radiation.

The magnetic field, about twenty times stronger than Earth's, is anchored to the core of the planet, and is tilted at an angle of 10°, so that the planet's rotation whirls the whole assemblage around with it at tremendous speed. On the side towards the Sun, there is a shock wave and a sharp boundary between the solar wind and the jovian 'magnetosphere', which fluctuates according to the pressure of the solar wind. The four Galilean moons all orbit within the magnetosphere, but the peak density of electrons and protons is close to the planet, below the orbit of Io.

The radiation belts emit synchrotron radiation continuously, at radio wavelengths shorter than one metre. But the most intense radio emission comes in bursts at longer (decametric) wavelengths. These bursts are triggered by Io. As the magnetic field is continually whipped past Io by the planet's rotation, it accelerates electrons in the vicinity of Io; the radio bursts probably arise as these electrons plunge into the ionosphere of Jupiter. Moreover, because Io conducts electricity, electrons and protons bouncing to and fro are channelled towards it along the magnetic field lines and form a tube of current through it (the Io flux tube). The heating from this current may contribute to the volcanic activity of the moon.

Io in turn contributes to the radiation belts, as atoms of oxygen, sulphur and sodium are scoured off Io's surface to form a diffuse ring close to its orbit (the Io plasma torus). The faint fluorescence of this torus has been recorded by very sensitive cameras on Earth. The emission from sodium (at the same wavelengths emitted by terrestrial streetlights) reveals neutral atoms close to Io, while the emission from sulphur reveals ions in the torus, which is locked to the tilted magnetic field.

On the side of Jupiter away from the Sun, the magnetic field is drawn out by the solar wind into a long comet-like tail waving in the breeze. The Voyager space-craft detected this tail as far out as the orbit of the next giant planet, Saturn.

4

Saturn encountered

RICHARD McKIM

Setting the scene

To the naked eye, Saturn is a yellowish-white star, with a mean opposition magnitude of
+0.7. It was the outermost known planet until 1781, and has always been a favourite
telescopic object. Good binoculars will currently show it as an elliptical disc – the form
of the rings projected upon the sky – but a powerful telescope will be needed for a better
view. Although easy to *look* at with a suitable aperture, Saturn is very difficult to *observe*.
It does not present rapidly changing global features like Jupiter, which hold the observer's
attention. At 9.5 AU from the Sun, Saturn also presents a smaller telescopic image, but
the amateur astronomer still has a part to play in the study of this planet.

Telescopically, Saturn shows a flattened yellow disc crossed by darker, warm-coloured
belts lying parallel to the planet's equator. At some point the rings pass in front of and
behind the globe. The disc is even more flattened than Jupiter's, with the equatorial
diameter standing at 120 000 km. The configuration of the rings passes through a $29\frac{1}{2}$-
year cycle; the last edgewise phase was in 1979–80, and at the time of writing (1990) the
rings are fully open on their north face. The rings vary in brightness from one ring to the
next and display several divisions. They cast their shadows on the ball of Saturn, which
in turn throws a shadow across the rings. Eight of Saturn's moons can be seen in a 30 cm
telescope, while Titan reveals its disc when viewed with larger apertures.

No Earth-based views, however, can approach the superb images returned by the
Voyager space-craft in 1980 and 1981. (Saturn was also visited by Pioneer 11 in 1979.)

1. Saturn drawn by Richard McKim, observed from the Pic du Midi Observatory, 20 July, 1986, at 20 h 20 m, using a magnification of × 533 in near-perfect seeing with a 1-metre Cassegrain reflector. Cassini's division, in black, is easily visible all around the ring system. Several other ring divisions, and many belts on the globe, can be seen.

Not only did these reveal the fine structure of Saturn's clouds and rings, but they added another seven moons to the total of ten or eleven detected during three centuries of ground-based astronomy.

The globe of Saturn

As with Jupiter, most of our pre-space age ground-based data about the rotation periods of Saturn's belts and zones had been gathered by amateur observers. The nomenclature for these features is similar to that in use for Jupiter. Occasional dark spots in the belts and white spots in the zones had shown us there exists a range of about 25 minutes between equator and poles. For example, a white spot in the North North Temperate Zone in 1960 had a period of 10 hours, 39 minutes, 9 seconds, while the famous equatorial white spot of 1933 had a period of about 10 hours 14 minutes. (The 1933 spot was discovered by UK amateur astronomer Will (W.T.) Hay, who financed his hobby by a professional career as a stage and screen comedian.) This range in rotation periods is some four times greater than for Jupiter; in modern terminology we say that Saturn has the more marked equatorial 'jet' of the two giant planets. In the early 1980s there were several dark spots visible on the south edge of the North Equatorial Belt. These gave a period similar to that of Hay's spot, but on the whole the belts and zones have recently shown only a delicate mottling with no large-scale features. Saturnian Spots are 'weather' features like those on Jupiter. Dark spots probably lie deeper in the atmosphere than bright spots, but have similar morphology. A number of spots were seen to be rotating clockwise (anticyclonically), revealing them to lie in high-pressure regions.

The most exciting spot activity of recent years occurred in 1990. On 25 September, an American amateur astronomer, Stuart Wilber, found a large white spot in Saturn's Equatorial Zone, similar to Hay's Spot of 1933. The spot expanded rapidly in length, especially in the preceding direction, and eventually most of the zone was filled with bright clouds. A preliminary analysis by the writer of the observations by members of the BAA Saturn Section showed that the centre of the spot had a rotation period of 10 hours, 13 minutes, 48 seconds between September and November. Images by the Hubble Space Telescope revealed that the spot overlapped the south edge of the North Equatorial Belt in a complex pattern of eddies and white rifts.

Voyager was able to refine our knowledge of Saturn's atmosphere, but not until the imaging team had greatly enhanced the contrast of the belts and zones by means of computer-processing and false colour techniques. It was found that Saturn displays features similar to Jupiter but they are of a smaller scale and muted by a thicker layer of overlying haze. Interesting features tracked by the space-craft cameras included a small red spot, since named Anne's Spot, at latitude 55° S, and a ribbon-like wave disturbance in the same region. Complex convective patterns are revealed by the Voyager images. Research by the BAA Saturn Section demonstrates that the positions, brightnesses and colours of the

2. *Beautiful Saturn with its ring system and three of its moons, Tethys, Dione and Rhea.*

Richard McKim

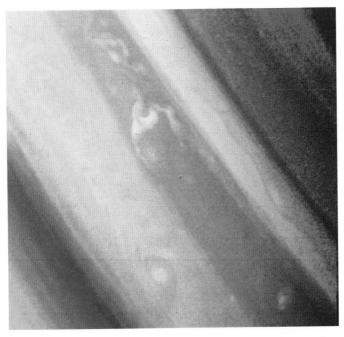

3. *This enhanced image of Saturn's cloud system was taken from a distance of 9 million km on 5 November, 1980.*

saturnian belts and zones are subject to change, sometimes in a seasonal manner. For this reason, further study is needed as Voyager could only yield a fingersnap view of the planet at one instant in its seasonal cycle.

The atmospheric envelope of Saturn, whose outer skin is the only part we can observe directly, is composed largely of material of solar composition: 93% hydrogen, 7% helium, with traces of other gases. Many gases which are present on Jupiter are hard to detect on Saturn, for they are frozen out at the lower temperature prevailing. Methane, ethane, ethyne, phosphine and ammonia are present, and so almost certainly are water, hydrogen sulphide, ethene and hydrogen cyanide. Photochemical reactions and atmospheric absorption account for the various colours we see. Yellow, orange and red tones can be produced by the photo-dissociation of phosphine to yield an aerosol of red phosphorus, while reactions between hydrogen sulphide and ammonia yield complex coloured polysulphides. Organic chromophores are probably also present in the atmospheres of both Saturn and Titan.

A good deal can be said of the probable structure of Saturn's atmosphere and interior. The atmosphere appears to be about 30 000 km deep, with the pressure increasing from about one terrestrial atmosphere on the outside to some two–four million atmospheres near the bottom. At such high pressures the hydrogen molecules split up into atoms and the hydrogen becomes 'metallic' and conducting. At very low temperatures helium droplets begin to condense in the metallic hydrogen layer and fall towards the centre of Saturn. As they fall inwards, gravitational potential energy is transformed into heat. This

55

process is believed to account for Saturn's 'internal heat source'; the planet radiates 2.8 times as much energy as it receives from the Sun.

Beneath the metallic hydrogen layer there is believed to be a rocky core containing metal oxides, metal silicates, and ices of water and ammonia, extending to about one-quarter of Saturn's radius. Needless to say, various 'models' fit the available data. All of these account for Saturn's extraordinarily low density of 0.70. Saturn is the only planet lighter than the same volume of water!

The magnetosphere

Saturn's magnetic field is less extensive than Jupiter's, but more extensive than the Earth's. The axis of rotation coincides closely with the direction of the magnetic field. Inside the magnetopause, the magnetosphere of Saturn is complex in structure, consisting of an outer toroidal ring of neutral and ionised hydrogen, inside which there is a smaller torus of energetic ions and finally an inner 'plasma' torus consisting mainly of oxygen ions. The planet emits a good deal of radio static in the 50 kHz–1 MHz range, due to electrons (probably from the solar wind) being accelerated along the magnetic flux lines towards the saturnian poles. The radio emissions follow a cycle of 10 hours, 39 minutes, 24 ± 7 seconds. Astronomers assume this to be the period of the core of the planet where the magnetic field originates. The period compares with those for atmospheric features in Saturn's temperature and polar latitudes.

The rings

The wonderful ring system, observed as such by the Dutchman Christiaan Huygens in 1655, is very complex. From Earth, three bright rings are visible, and one obvious division (Cassini's division) separates the outer ring (A) from the inner rings (B and C). In good seeing with large telescopes the rings resolve themselves into many fine divisions. From Earth, a faint ring exterior to ring A was detected, while in 1969 Pierre Guérin described the faint 'D' ring, between C and the globe. This was a controversial discovery, but UK astronomer John Murray confirmed it in 1971.

In addition to confirming these difficult telescopic discoveries, Voyager showed that there are many fine ring subdivisions or gaps in the system: over 1000 can be traced in the images! Stellar occultation data from Earth had shown that the rings must be more complex than they appeared visually, but no one had imagined that the increased resolution of Voyager would show such a multitude of ringlets.

Long before the space age it was known that the rings must consist of a swarm of small particles. Kepler's laws show that the period of rotation of a ring particle, planet or satellite lengthens with increasing distance from the primary body. Thus the rings cannot consist of a continuous sheet of matter for they would be torn apart by the differential rotation. The French mathematician Eduoard Roche showed in 1850 that no large satellite could approach to within 140 000 km of the centre of Saturn without being torn apart by tidal forces. This distance is now known as Roche's Limit, and the 'bright' rings lie within this boundary. James E. Keeler in 1895 obtained observational proof of the rotation of

4. As Voyager 1 passed beneath the ring plane on 12 November, 1980 it recorded this view of Saturn's rings. The image has been enhanced to bring out details of the ring system. The broad bright bands near the centre are the C ring. Reddish-brown denotes the B ring. It appears dark because it contains too many particles to allow the sunlight to scatter all the way through.

the rings from the Doppler effect detectable in their spectra taken at Lick Observatory. It was also known that the rings must be very thin. William Herschel (1789) thought the thickness could not exceed some few hundreds of kilometres, and his estimate has been reduced to about 1 km or less by the space-craft data. Viewed edge-on from Earth, the rings all but disappear.

Voyager gave us much new data about ring-particle sizes and ring structure. Let us consider the large-scale ring structure first. Most ring divisions are fine gaps. Cassini's division is not free from matter (already known from Earth-based data, which showed a

5. *Voyager 2 recorded this image of Saturn's rings on 22 August, 1981. 'Spoke' features are visible in the B ring.*

brightening of the division under oblique lighting), and has several ringlets within it. Before 1980 it was thought that the various divisions were due to orbital resonances with the different satellites. Particles straying into a gap would be removed by the gravitational influence of a particular moon. But these 'Kirkwood Gaps' (1867) cannot account for the fine structure which exists at close range, so that collisional (or dynamical) processes must also play a part. One additional factor is the involvement of 'shepherd' satellites – another Voyager discovery. Pioneer 11 found a fine ring just exterior to ring A, named the F ring. Voyager 1 found two tiny moonlets orbiting either side of it; there can be no doubt these bodies have the task of marshalling any straying ring particles back into line! The F ring, which was found to consist of three components, was braided like the strands of a rope, owing to the influence of the shepherds. Another moonlet, discovered just outside the A ring is thought to prevent the spreading of the bright rings.

A sensation was caused by the discovery of dark, radial 'spokes' in the rings, radiating outwards from the inner edge of ring B. (Earth-based observers had sometimes seen them in ring A.) They rotated with the ring, and are presently thought to be due to tiny, micron-sized particles levitated above the ring-plane by electrostatic forces. In forward-scattered light, on Voyager's approach, they look dark, but in back-scattering (as the space-craft departed) they are bright. The spokes thus contain more fine dust than their surroundings. The spokes must represent some interaction between the ring particles and Saturn's extensive magnetosphere which normally extends beyond the orbit of Titan, but which varies with the solar wind flux.

What about the sizes of the ring particles? Space-craft data concerning relative brightnesses of the rings coupled with solar occultation observations, over a broad wavelength band, showed that a whole spectrum of particle sizes was present. Most particles are 1–5 cm in radius, but all sizes up to about 10 km radius are present at some level. Size distribution is different for each ring. The spokes and the F ring have many tiny particles, but ring C is likely to be more sparsely populated by larger boulders. Furthermore, the particles have different albedos. They are predominantly icy, or ice-covered, but some are darker and largely rocky, silicate-based conglomerates.

Amongst the moons

Saturn has the largest number of moons of any known planet, with seventeen satellites now recognised. The first nine of these were discovered by 1898, mostly by visual observation. They range in size from small icy bodies a few tens of kilometres in diameter up to Titan, which is of planetary dimensions.

Titan is the largest and most interesting saturnian satellite, and possesses an extensive atmosphere with an estimated ground pressure comparable to that of Earth. With a diameter of 5150 km it is a little larger than the planet Mercury, but has a lower mass on account of its much lower density. Like the other moons of Saturn's system, its rotation is captured. That is, it always keeps the same face turned towards Saturn, as a result of

6. *Saturn from a distance of 13 million km. Two of its moons, Tethys (above) and Dione, have also been caught in the image. Dark shadows cast by Tethys and the planet's rings are visible against the bright disc of the planet.*

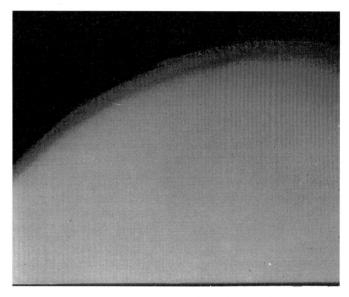

7. This view of Saturn's moon, Titan, shows the thick atmosphere that surrounds it.

the action of tidal friction over many millions of years. Titan's 400 km thick atmosphere is composed largely of nitrogen, but contains significant amounts of methane, other hydrocarbons and also cyanogen and hydrogen cyanide. The surface temperature is estimated at 92 K and there may be liquid methane rain and seas, with liquid nitrogen at the poles! The bizarre orange colour of Titan, ascribed to a photochemical smog, has been the subject of much speculation. The Voyager cameras could not penetrate the smog layers, merely recording low-contrast bands in the atmosphere.

All the other satellites are small, icy, airless worlds. Most are densely cratered, Rhea (diameter 1530 km) having the greatest cratering density. But Enceladus (500 km in diameter) shows little such detail, and must have been extensively resurfaced after the impact bombardment which cratered the satellites in early solar system history. Geological activity is in evidence on other satellites too – such as Tethys and Dione, where grooves (or sinuous valleys) are apparent.

Mimas, a small circular body 392 km across, shows an enormous impact crater ('Herschel') about one-third of the diameter of the moon itself. It is thought that the impact of a planetoid only marginally larger than the one which caused this crater would have completely shattered Mimas. This satellite is one of the smaller moons, and orbits Saturn just within the outer limit of the tenuous E ring.

Iapetus (diameter 1460 km) was not imaged at high resolution, but appears to be a very unusual body from the encounter images. Earth-based observations since its discovery by J. D. Cassini in 1671 have shown that one hemisphere of Iapetus is much darker than the other, so that the satellite is always significantly brighter at western elongations. Voyager images show very dark material on one side of the moon, giving that side only 20% of the brightness of the other side. Its nature remains a mystery.

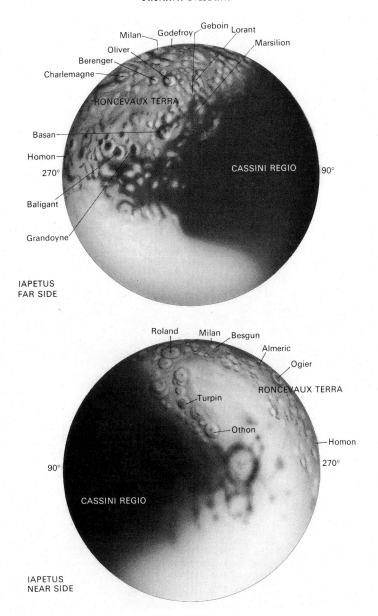

8. *One hemisphere of Iapetus is significantly darker than the other. The main features of the moon are shown on these maps.*

9. Dione's cratered surface is clearly visible in this Voyager 1 image. Central peaks can also be seen in the largest of the craters which measure up to 96 km in diameter.

As mentioned earlier, several new moons were found on the space-craft images. Apart from the 'shepherds', another small moon is in the same orbit as Tethys, and is therefore co-orbital. Doubtless there are other smaller bodies which were not seen by the cameras! However, one must also regard the rings as an uncountable collection of tiny satellites, at the other end of the spectrum of particle size.

The future

Professional astronomers continue to investigate their models of Saturn's atmosphere and interior, and the dynamics of the rings and satellites. There is also much research to be done concerning the crustal evolution of the moons.

The amateur continues to monitor Saturn for two reasons. Firstly, he or she may discover a new spot outbreak, and these unpredictable eruptions present the best chance of adding to our knowledge of the movements of features in Saturn's visible atmosphere. Secondly, the amateur is well-equipped to study long-term changes in the visibility and brightness of the belts and zones.

Telescopically, Saturn with its solar system in miniature will always be an object of indescribable beauty.

5

Uranus, Neptune and Pluto

PATRICK MOORE

Saturn was the outermost planet known in ancient times. Then, in 1781, William Herschel discovered Uranus during a systematic 'review of the heavens' carried out with a home-made telescope. Irregularities in the movements of Uranus led to the tracking-down of Neptune, in 1846. The actual discovery was made by Johann Galle and Heinrich D'Arrest, at the Berlin Observatory, working on calculations sent to them by the French mathematician Urbain Le Verrier.

From the outset it was clear that, though Uranus and Neptune are giant planets, they are very different from the Jupiter/Saturn pair, and also different from each other in many respects, even though in size and mass they are almost twins.

Uranus is generally described as greenish and Neptune as blue, but surface details are almost impossible to make out from Earth. It was found that Uranus, unlike the other three giant planets, lacks a strong internal heat-source, and it also has an extraordinary axial inclination; the tilt amounts to more than a right angle, so that each pole has a 'midnight sun' lasting for twenty-one Earth years, with a corresponding period of darkness at the opposite pole. The reason for this strange state of affairs is unknown. The suggestion that in its early history Uranus was struck by a massive body, and literally knocked sideways, does not seem very plausible, but it is hard to think of any other explanation.

In March, 1977, an occultation of a star by Uranus was observed. Both before and after occultation the star 'winked', betraying the presence of a system of thin, dark rings; the

	Uranus	Neptune	Pluto
Mean distance from the Sun, millions of kilometres	2869.6	4496.7	5900
Revolution period, years	84.0	164.8	247.7
Rotation period	17h 14m	16h 3m	6d 9h 17m
Axial inclination, degrees	98	28.8	118
Equatorial diameter, km	51 118	50 538	2445.4
Mass, Earth = 1	14	17	0.0025
Orbital eccentricity	0.047	0.009	0.248
Specific gravity	1.27	1.77	1.4
Maximum magnitude, seen from Earth	+ 5.6	+ 7.7	
Maximum apparent diameter, seen from Earth	3".7	2".2	

ring-system was subsequently confirmed by further occultations, and was photographed in infra-red. Occultation results for Neptune were inconclusive, and it was suggested that there might be 'ring arcs'. Telescopic observation showed that Uranus had five satellites and Neptune two.

Such was the state of our knowledge when Voyager 2 was launched in 1977. The space-craft had an ambitious programme. A fortuitous alignment of the four giant planets meant that Voyager could by-pass all of them in turn; Jupiter in 1979, Saturn in 1981, Uranus in 1986 and Neptune in 1989. Virtually all our detailed knowledge of Uranus and Neptune has been obtained from this one probe.

In 1986, as Voyager 2 approached Uranus, the planet's south pole* was facing the Sun and the Earth, and was in the middle of its long 'summer', so that the approach was pole-

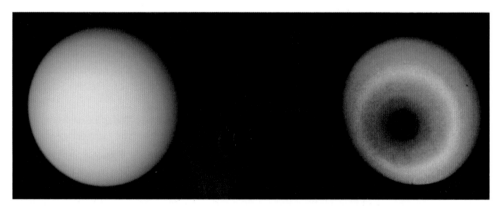

1. These two views of Uranus were produced from Voyager 2 images recorded on 10 January, 1986, from a distance of 11 million miles. At left is Uranus as human eyes would see it, at right is a false-colour image which shows detail in the polar region of the atmosphere.

---

* There is disagreement as to the proper nomenclature for Uranus's poles. The International Astronomical Union convention is that a pole above the ecliptic is 'north', and in this case the pole sunlit during the Voyager pass is 'south'. However, the Voyager team uses an alternative convention, making the sunlit pole 'north'. In this chapter I have followed the IAU convention.

2. *The Uranus ring system imaged by Voyager 2 while in the shadow of the planet. All the previously known rings are visible as well as bright dust lanes not previously seen. The 96-second exposure produced the streaks due to trailed stars.*

on; for this reason alone the view was very different from that at Jupiter or Saturn. Moreover, there was very little surface detail. A few clouds were seen, but there were no vivid colours, and the general appearance was remarkably bland. However, it was established that different latitudes have different rotation periods, and this means that there are strong winds. It was also found that the temperatures at the poles and at the equator were much the same.

It seems that Uranus contains more 'heavy elements' than Jupiter or Saturn. There is presumably a solid, rocky core, surrounded by a deep layer in which ammonia, water and methane are mixed as 'ices'; in the atmosphere, water, ammonia and methane condense in that order to form cloud-layers. Methane freezes at the lowest temperature, and so forms the top layer, above which comes the predominantly hydrogen atmosphere containing appreciable quantities of helium and other gases such as neon.

It had been expected that Uranus would be a radio source, and would have a magnetic field. This proved to be correct, but there was one major surprise. The magnetic axis is nowhere near the axis of rotation, but is inclined to it at an angle of almost 60 degrees; it is also reversed with respect to that of the Earth, and to complete the picture the magnetic axis is considerably offset from the centre of the globe. The uranian magnetosphere extends to 590 000 km on the day side of the planet and 6 000 000 km on the night side. Ultra-violet observations from Voyager showed that there are strong emissions on the day side, producing what has become known as the electroglow. It may be caused by electrons

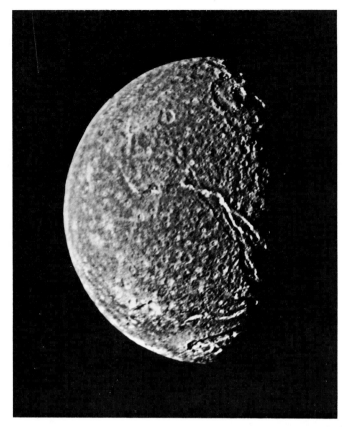

3. *This picture of Titania is a composite of two images taken by Voyager 2 on 24 January, 1986. Impact craters and fault valleys cover the surface. A 200 km diameter crater cut by a fault valley more than 96 km wide is at the very bottom of the disc.*

exciting hydrogen molecules in the upper atmosphere, though it is not clear just how these electrons obtain all their energy.

The rings also provided surprises. Ten are now known, all of which are dark; they seem to be made up mainly of particles a few metres in diameter, and not all are circular. There is also a considerable amount of 'dust'. The outermost ring, known as the ε-ring, has a width of between 22 and 93 km, and is inclined to the uranian equator at an angle of almost 8 degrees; all the other rings lie virtually in the equatorial plane.

Ten new satellites were discovered by Voyager 2, all closer in than Miranda, the innermost of the previously-known satellites. All ten of them are small; even the largest of them, now named Puck, is no more than 154 km in diameter. However, the larger satellites provided much of interest. Oberon and Titania are icy and cratered; Umbriel has a darker, more subdued surface with numbers of craters; while Ariel is dominated by craters and by broad, branching, smooth-floored valleys which look as though they have been cut by liquid – though Ariel is much too insignificant to retain any sort of atmosphere.

Miranda is completely different. Though a mere 472 km in diameter, it has an amazingly varied surface, with several distinct types of terrain. There are old, cratered plains, brighter areas with cliffs and scarps, and 'ovoids', large trapezoidal-shaped regions 200 to 300 km across. The ice-cliffs may be as much as 20 km high. It has been suggested that Miranda may have been broken up by a colossal impact and subsequently re-formed, but this is no more than a theory, and will be very difficult to prove.

Neptune, moving far beyond the orbit of Uranus, was the last target of Voyager 2. Success was by no means guaranteed; after all, Voyager had been planned, built and launched in the 1970s, so that it qualified as an old space-craft. Yet during the Neptune pass it performed faultlessly. After a journey of more than four and a half thousand million kilometres, lasting for twelve years, Voyager 2 reached its target within six minutes of the planned time. On 25 August, 1989, it skimmed within 4900 km of the cloud-tops over Neptune's darkened north pole. At that time the distance from Earth was 4425 million kilometres, so that the radio signals took 4 hours 6 minutes to reach Mission Headquarters at the Jet Propulsion Laboratory in California.

Of the two known satellites, Triton was regarded as very much of an enigma. It was thought to be at least as large as our Moon, but is unique among large planetary satellites in having retrograde motion; there were suggestions that, like Titan in Saturn's system,

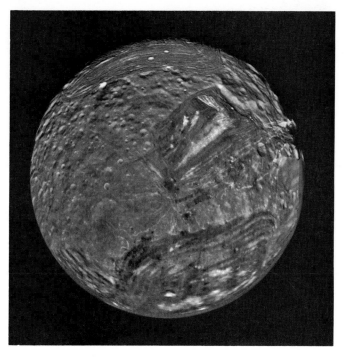

4. Nine images were combined to produce this full-disc view of Uranus's moon Miranda. Seven high-resolution images from Voyager 2's closest approach make up the major portion of the image, more distant, lower resolution images, complete the gaps along the limb.

5. *Voyager 2's narrow-angle camera recorded approximately two and a half rotations of Neptune. This picture from the sequence shows two of the four cloud features tracked by Voyager. The large dark oval near the western limb (at left) has bright clouds immediately to the south and east of it. The second dark spot is seen at the lower right edge.*

it might have an atmosphere dense enough to conceal its true surface. Nereid's diameter is much less, but it too is unusual in having a very eccentric orbit, more like that of a comet than a satellite.

Unfortunately, Nereid was in the wrong part of its path during the Voyager mission, and no detailed images of it were obtained, but during the approach six new inner satellites were discovered, one of which was actually larger than Nereid; a good image of it was obtained, showing a darkish, cratered surface. All these new satellites are so close to Neptune that they are quite undetectable from Earth.

Ring-arcs had been expected, from the occultation results. Voyager found that there are in fact complete rings — three in number, plus one broad 'plateau' of particles. The outer ring has most of its material clumped into three definite regions, and it was these which had been detected from Earth, giving an impression of disconnected arcs. The faintest of the complete rings was only just above Voyager's threshold of visibility.

Voyager showed definite details on Neptune; revealing much more to see than had been the case with Uranus. As the space-craft moved in, more and more features on the disc came into view. Radio emissions were also detected, and these enabled the rotation period of the planet itself to be fixed at 16 hours 3 minutes, rather shorter than had been anticipated. Next came the entry into the neptunian magnetosphere — and another major surprise.

Neptune has a magnetic field, rather weaker than those of the other giant planets. The unexpected discovery was that the magnetic axis is inclined to the rotational axis by a full 50 degrees. Magnetically, in fact, Neptune is very like Uranus, and quite unlike Jupiter or Saturn; as with Uranus, the magnetic axis is considerably offset from the centre of the globe. It has been suggested that the offset magnetic axis of Uranus was in some way associated with the 98-degree tilt of the rotational axis, but this does not apply to Neptune. It had also been suggested that we had happened to catch Uranus at the time of a 'magnetic reversal', but to believe that Uranus and Neptune were experiencing similar reversals at the same time would be too much of a coincidence. At the moment, we have to admit that we have no satisfactory explanation.

It may well be that the 'dynamo' electrical currents responsible for Neptune's magnetic field are close to the surface of the planet, rather than concentrated near the core as with the Earth. The field strength at the surface ranges from a maximum of 1.2 gauss in the southern hemisphere to only 0.06 gauss in the northern. (At the Earth's equator, the field strength is approximately 0.3 gauss.) Weak aurorae were recorded on Neptune — but, of course, these occur near the magnetic poles, which are a long way from the poles of rotation.

The most prominent feature on Neptune is the Great Dark Spot (GDS) — a vast, circulating storm almost as large as the entire Earth. It lies at a latitude of 22 degrees south, and has a rotation period of 18.3 hours. Above it are bright cirrus-type clouds, composed of methane ice crystals; these clouds lie at least 40 km above the main cloud-deck, and below them is a layer which seems to be more or less clear. The cirrus clouds change so rapidly that at first the investigators found it hard to identify them from one neptunian 'day' to another.

Other features were seen, mainly in the southern hemisphere (remember, it is the south pole which is having its long summer; during the Voyager fly-by, the north pole was in darkness). A second dark spot, at latitude 54°S, has a rotation period of just over 16 hours, almost the same as that of the planet's core. At an intermediate latitude is a roughly triangular, bright cloud feature which has been nicknamed the 'Scooter' because of its faster rotation period. Evidently it lies at a lower level.

From these various measurements, it became possible to work out a wind pattern for Neptune, and again the results were decidedly unexpected. Near the equator, the wind is westerly or retrograde, i.e. opposite to the east-to-west direction of the rotation of the planet itself. The speed is around 250 m/s relative to the core. At the latitude of the GDS,

6. *This image of Neptune's rings was taken by Voyager 2 from a distance of 1 100 000 km. It was recorded as the space-craft travelled away from the planet and was the first to show the rings in detail.*

the westerly windspeed reaches some 325 m/s; further south it decreases, falling almost to zero at latitude 50° S, and closer to the pole there is an easterly wind, reaching about 50 m/s at latitude 60° S and then falling to zero again at the pole.

High-resolution images confirmed that some of the cirrus-type clouds are very high, and may reach an altitude of 100 km above the main deck. One particularly beautiful picture showed shadows cast on to the globe − a feature not seen on any of the other giant planets.

Methane is all-important on Neptune, even though it makes up only about 2% of the atmosphere; of the rest, 85% is hydrogen and 13% helium, while the darker cloud features, well below the main methane cloud deck, may contain large amounts of hydrogen sulphide. There seems to be a well-defined methane cycle: (1) Solar ultra-violet radiation destroys methane in the high atmosphere by converting it to hydrocarbons, such as ethane and acetylene. (2) The ethane and acetylene descend into the colder, lower stratosphere, where they condense into solid particles. (3) These particles fall into the

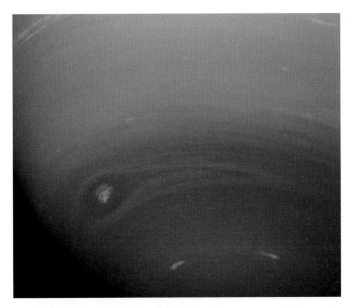

7. Neptune's southern hemisphere. The almond-shaped structure at the left is a large cloud system. The smallest features visible are around 40 km across.

warmer troposphere, evaporate, and are converted back into methane. (4) Buoyant, convective methane clouds then rise to great heights, returning methane vapour to the stratosphere. Therefore, the complete cycle means that there is no net loss of methane.

After skimming over Neptune's north pole, Voyager 2 had to cross the ring-plane, and this caused a certain amount of trepidation. Between fifteen and ten minutes before the actual crossing, the rate of impact with ring particles increased alarmingly, finally peaking at around 300 per second; when it declined steeply after the moment of crossing, there was a sigh of relief at Mission Control! A few hours later, Voyager passed within 38 000 km of Triton, and provided what most people regarded as the highlight of the whole encounter.

It was soon obvious that Triton's atmosphere was much too thin to hide its surface. The ground pressure is no more than 10 microbars – enough to produce some limb haze, but nothing more; most of it is nitrogen, together with some methane. The diameter was found to be only 2720 km, smaller than our Moon and much smaller than had been expected. This meant that it had a high albedo, and was very cold indeed. The specific gravity is 2.02, so that presumably the globe is made up of roughly one-third water ice and two-thirds rock.

As with Neptune, it is Triton's southern hemisphere which is now experiencing summer, though the seasons are much more complicated than with Neptune itself. Voyager showed a large southern cap of 'pink snow', composed of nitrogen ice with an admixture of methane ice. The northern hemisphere is different in colour and in surface detail; it is bluish, with numerous shallow valleys. There is little surface relief, and there are very few

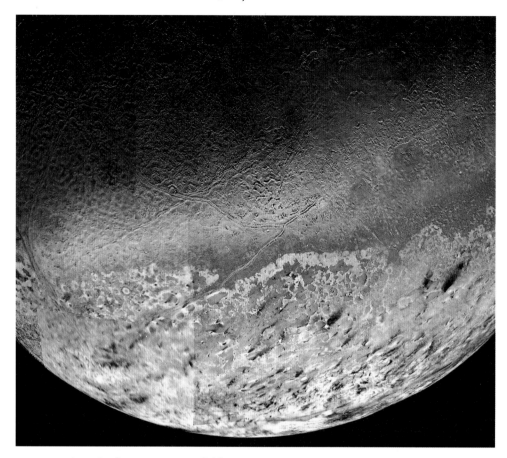

8. *Around a dozen images, recorded by Voyager 2 on 25 August, 1989, were combined to produce this view of Neptune's large satellite Triton.*

impact craters, but there are large enclosures which give the impression of being frozen lakes. The boundary between these two hemispheres is very marked.

It also seems that activity may be going on at the present time. If there is nitrogen ice on the surface, then presumably there must be a layer of liquid nitrogen only a few metres below. If this percolates upward, it will reach a region where the pressure is insufficient to keep it in liquid form – and it will burst forth, as a geyser of nitrogen ice and vapour; the deposit will be carried along by wind in the tenuous atmosphere to produce a dark streak. Numbers of these streaks have been found. Nothing of this sort had been anticipated; Triton is unlike any other world in the solar system.

All the evidence now supports the theory that Triton is not a true satellite of Neptune, but was once an independent body which was captured in the remote past. This would account for its unique retrograde motion. Its first path round Neptune would have been eccentric, as Nereid's is today, but the orbit would gradually be forced into the circular form; in the process there would be tidal flexing and heating, so that water-dominated material would pour out on to the surface and flood wide areas, as we see today in the northern hemisphere.

As Voyager 2 drew away, it took one more picture, showing Neptune and Triton as neighbouring crescents. Before long it had receded so far that details were lost – and Voyager had started its never-ending journey out of the solar system.

It is a pity that Voyager could not be sent on to an encounter with Pluto, the one remaining member of the planetary family. Unfortunately, Pluto was in the wrong part of its orbit. It has an eccentric path, and near perihelion it is closer in than Neptune; perihelion occurred in September, 1989, so that between 1979 and 1999 it is Neptune, not Pluto, which ranks as 'the outermost planet'. To send Voyager on to Pluto would have meant burrowing deep into Neptune's globe, which was out of the question for obvious reasons.

Pluto was discovered in 1930 as the result of a deliberate search by Clyde Tombaugh at the Lowell Observatory, Flagstaff. Calculations made many years earlier by Lowell himself, on the perturbations of Neptune and (particularly) Uranus, had indicated that a ninth planet should exist, and indeed after its discovery, not so very far from the calculated

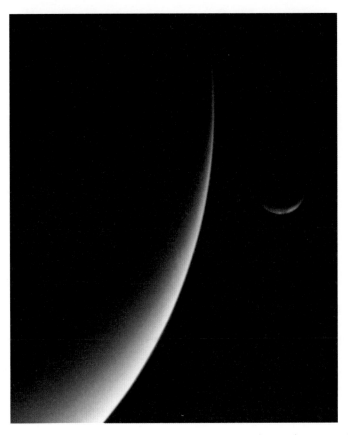

9. *The crescents of Neptune and Triton imaged by Voyager 2 three and a quarter days after closest approach to Neptune.*

position, its image was found on plates taken at Flagstaff during Lowell's lifetime and also by Milton Humason at Mount Wilson in 1919. However, is soon became clear that Pluto could not exert measurable perturbations upon the movements of giant planets, because it has too low a mass. Its diameter is a mere 2445 km, smaller than our Moon, with a mass only 18.5% that of the Moon. Therefore, in view of its insignificance and its curious orbit, should it be classed as a true planet?

It has a very small apparent diameter, but fortunately it has been found to be accompanied by a second body, Charon, which was discovered by J.W. Christy in 1977 from photographs taken at the US Naval Observatory for position purposes. Charon has a diameter of 1930 km, so that it is comparable with Pluto itself; the two are only

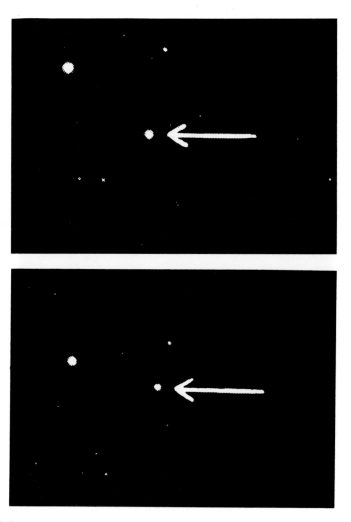

10. *Clyde Tombaugh's photographic plates of Pluto. It was the movement of the planet (arrowed) against the background of the stars that led to Pluto's discovery.*

19 000 km apart, and are 'locked' so that Charon's revolution period is equal to the rotation period of Pluto: 6.39 Earth-days.

The discovery of Charon meant that for the first time a really reliable mass of Pluto could be calculated (or, to be more precise, the combined mass of the Pluto/Charon system). Moreover, by a lucky chance, the two were so oriented during the mid and late 1980s that they periodically occulted each other, making it possible to study them separately. Methane ice on Pluto had been detected spectroscopically in 1977, and the occultation techniques established that the methane ice mantle covers the planet's poles, with a 'warmed' band across the equator; the coating of Charon appears to be water ice. Pluto has an extensive though very tenuous atmosphere of methane, though when near aphelion it may be that the atmosphere freezes out, so that for part of its 247.7-year long revolution period there is no atmosphere at all. No atmosphere of any kind has been detected with Charon.

Pluto is roughly twice as dense as water, so that rock seems to make up from 68 to 80% of its mass. The core rock is presumably hydrated; it may be either that Pluto is undifferentiated, or else that it has a silicate core surrounded by a mantle of ice 200 to 300 km thick.

Certainly Pluto is an enigma. It is slightly smaller than Triton, though earlier suggestions that it used to be a member of Neptune's satellite system have been generally discounted. We must also consider Chiron, discovered by C. Kowal in 1977, which spends most of its time between the orbits of Saturn and Uranus, and has been given an asteroidal number, 2060, even though it has developed a coma and may possibly be nothing more nor less than a giant comet. Is the Pluto/Charon pair merely the brightest of a whole swarm of similar objects in the far reaches of the solar system? Has it any cometary associations? Is it merely a remote asteroid pair, or should it after all be given planetary status? At the moment we do not know, and no probes are as yet scheduled to go to it and find out.

Whether or not another planet, further from the Sun, awaits discovery is an open question. There are various indications that 'Planet Ten' exists, because of small unexplained irregularities in the movements of Uranus and Neptune, but we have no real idea of where it is, or even definite proof that it exists at all. In this quest we cannot expect help from the Voyagers, or indeed from Pioneers 10 and 11, which are now on their way out of the solar system.

If all goes well, Voyager 2 will remain in touch with Earth until about the year 2020, by which time it will have approached the edge of the heliosphere – that is to say, the region of the Galaxy in which the Sun's influence is dominant. The same applies to Voyager 1. After that we will lose both space-probes, and they will go on their ways between the stars unseen, unheard and untrackable. It has been calculated that in about 290 000 years Voyager 2 will pass within 4.2 light-years of Sirius, always provided that it has not been destroyed by a collision with some wandering body.

Just in case the Voyagers are discovered and collected by some alien race, each carries a plaque, together with a record of 'Sounds of Earth' ranging from a crying baby to a pop group, a symphony, and a speech by the then President of the United States, Richard Nixon. Admittedly, the chances of collection are slim, but they are not nil, and any alien scientists would presumably be able to identify the place of origin of their interstellar visitors – provided, of course, that they had access to suitable record players!

Meanwhile, we must be proud of the achievements of Voyager 2. It has done all and more than its makers could have hoped, and it has increased our knowledge of the outer solar system beyond all recognition. Uranus and Neptune remain intriguing and in many ways mysterious, but at least we know far more about them now than we did less than a decade ago.

6

Comets and meteors

DAVID W. HUGHES

Introduction

Comets and meteors are closely related. As a comet travels through the inner solar system, it decays, and this loss of mass results in the release of prodigious amounts of gas and dust. The larger dust particles (10^{-6}–10^{5} g) are pushed away from the nucleus by gas pressure and are accelerated to velocities that are only a fraction of a kilometre per second. The cometary nucleus at the time has a heliocentric velocity of many tens of kilometres per second, so the dust particles find themselves on orbits that are nearly the same as the parent comet's. As time progresses they either slowly gain on or slowly fall behind the comet. After 1000 years or so, they have formed an annulus of dust around the cometary orbit, an annulus that is much thinner at perihelion than at aphelion. If the Earth happens to pass through this annulus, some of the dust particles collide with the upper atmosphere. There they burn up leaving behind them a fleeting trail of ionised and excited atoms and molecules that can be seen from the ground as a meteor.

Comets from the ground

Our knowledge, or lack of knowledge, about comets in the early 1990s can be appreciated by listing, and then trying to answer, a few questions. The questions might look simple, but we still have a long way to go before we will know the correct answers.

77

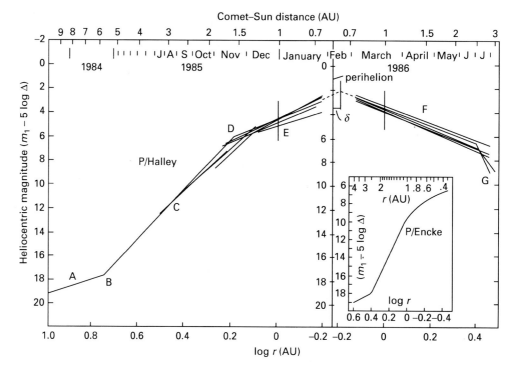

1. *The heliocentric apparent magnitude of Halley's Comet, (m_1-5 logΔ), where m_1 is the apparent magnitude of the comet as seen from Earth and Δ AU is the comet–Earth distance, is plotted as a function of the logarithm of the comet–Sun distance, r AU, for the time period 1984 to July 1986. Perihelion passage took place on 9 February, 1986. The individual lines on the diagram correspond to observations made by different observers. The insert shows the pre-perihelion light curve of another short-period comet, P/Encke, and it can be seen that there are distinct similarities. The bare nucleus phase is represented by A, and the heliocentric brightness in this region obeys an r^{-2} relationship. Sublimation starts around 6 AU and in phase C the comet obeys an $r^{-8.5}$ relationship. In the range $1.7 > r > 0.59$ AU the comet has a massive coma and the brightness follows an $r^{-3.3}$ relationship. Notice that the post-perihelion activity at a specific heliocentric distance is higher than the pre-perihelion activity.*

(i)   What is a comet?
(ii) Are they all the same?
(iii) How many comets exist in the solar system?
(iv) How do they decay?
(v)   Where do they come from?
(vi) How were they produced?
(vii) Will they *really* provide such a vital clue as to the mechanism
      responsible for the formation of the solar system?

Before we launch into this task, let us review the story so far.

We know that the fount of all the cometary activity is a kilometric-sized nucleus that is made up of snow and dust. If this nucleus comes to within about 2.5 AU of the Sun, the snow starts to sublimate and the gas that is produced quickly leaves the nucleus pushing away some of the dust as it goes. The nucleus soon surrounds itself with a large coma of gas and dust, and some of the smaller dust and the ionised gas is pushed away from the nucleus in the antisolar direction, forming a plasma tail and a dust tail.

To date, the earthbound observer has seen and recorded about 700 comets. This number is steadily increasing, and in 1989 thirty-four cometary apparitions were recorded, fourteen of which were new long-period comets. These long-period comets usually take well over 200 years to go around their orbit. Famous recent examples are Arend–Roland (1957 III), Ikeya–Seki (1965 VIII), Bennett (1970 II), Kohoutek (1973 XII), and West (1976 VI), the date indicating the year of perihelion passage and the roman numeral the order of their passage during that year. The orbits of long-period comets are inclined at random to the ecliptic plane. At their aphelia these comets can easily be 10 000–100 000 AU from the Sun. As comets spend a considerable proportion of their orbital period close to aphelion, this produces a concentration of comets in this region, known as the Oort–Öpik Cloud. There are about one million million comets in this cloud, and the cloud is rather susceptible to minor disruptions produced by passing stars and molecular clouds. On any of the occasions when a comet from the Oort–Öpik Cloud travels through the inner solar system, the gravitational influence of Jupiter (a planet that accounts for nearly 70% of the mass of the solar system) might slightly change the semi-major axis of the cometary orbit, either pulling the comet into the inner solar system or pushing it further out and even closer to passing stars. The average long-period comet is considerably brighter (by about five magnitudes) than the average short-period comet, although the nucleus is probably only slightly more massive. The long-period nucleus is much rougher, having a higher surface area to mass ratio and is also actively sublimating snow over a much larger percentage of its total surface area.

About 135 of the comets recorded so far have short periods. These comets have been captured into the inner solar system, and the mean period of the group is about 7.6 years and the mean inclination is about 11 degrees. Famous examples are P/Halley, which has a retrograde orbit and returns every 76 years, P/Encke, which has a period of 3.3 years and has been seen passing perihelion fifty-five times and P/Schwassmann–Wachmann 1, which orbits between Jupiter and Saturn with a period of around 16 years. (The P stands for periodic and the 1 indicates that Schwassmann and Wachmann discovered quite a few comets together, of which this is the first.) The orbits of short-period comets are precessing and changing slowly, again due to the gravitational influence of the outer planets. As Jupiter was the major capturer, these comets still tend to have aphelia close to the orbit of Jupiter, and a jovian flyby has a considerable potential for moving them into smaller or larger orbits. Short-period comets are only slightly less massive than the long-period ones, but they have decayed more and their nuclei are probably much smoother and also tend to be covered with large areas of insulating dust which restricts the active snowy regions to a few per cent of the surface.

As short-period comets pass perihelion relatively frequently, the effect of mass loss is more pronounced. The change in the absolute magnitude of a comet per orbit is $-0.724\Delta M/M$, where M is the pre-perihelion mass and ΔM is the mass lost during perihelion passage. (The absolute magnitude is the value the apparent magnitude would have if the comet was 1 AU from the Sun and was seen from a distance of 1 AU. For P/Halley, $\Delta M/M$ is typically 8×10^{-4} per apparition, so the comet has hardly got fainter over its observed thirty appearances, the absolute magnitude changing over this period by a negligible 0.02. The comet does, however, lose around 3×10^{14} g per apparition, and as much as 30% of this is in the form of large dust particles that feed the meteoroid stream that observers on Earth see as the Orionid and Eta Aquarid meteor showers each year.

The ground-based observations of comets can be crudely divided into three types: monitoring, imaging, and spectroscopy.

Monitoring

The monitoring of the overall activity of the comet using small telescopes and binoculars yields the way in which the cometary apparent magnitude varies as a function of time. The work of the dedicated amateur is vital in this field. A typical example of the results is shown in Figure 1. It can be seen that a comet goes through different stages of activity. When it is a long way from the Sun (say comet–Sun distance, r, exceeds 6 AU) the nucleus is so cold that it is completely inactive. Comets that never get closer to the Sun will remain in a deep-freeze state and will never lose mass. When comets are closer to the Sun than about 2 AU, the temperature of the nucleus varies very little, being of the order of the melting point of water ice. Energy absorbed from the solar radiation flux is mainly used to sublimate the snow, and very little goes towards increasing the temperature of the dusty nucleus surface. The typical accuracy of the apparent magnitude observations, m_1, that make up the graphs in Figure 1, is ± 0.3. The observations can usually be obtained only once a day, because the comet, being close to the Sun, only appears briefly in either the pre-dawn or post-sunset sky. Effects due to the gas jets from a spinning nucleus are thus ill-defined. Continuous observations of comets from space would be very useful here. Cometary nuclei are expected to be spinning with periods of around ten hours. The emission of gas and dust will vary in a patchy fashion over the nucleus surface, and this will cause the brightness of the inner coma to fluctuate with a period equal to the nucleus spin period. Unfortunately the spin mode will not be simple, and the spin axis is expected to be precessing wildly. If the mass loss from the nucleus is not the same in all directions, both the spin rate and the spin axis orientation will vary slightly from one perihelion passage to another. Gross asymmetries in the light curve, with respect to the time of perihelion passage, have been observed in certain comets, and these are probably due to the nucleus having a spin axis orientation that shields the active areas from the solar flux over large portions of the inner solar system orbit. The active areas can also change in size. They can decrease if insulating dust layers build up on the surface, or increase if these insulating layers break away. The absolute magnitude of a specific comet (the value of $m_1 - 5 \log \varDelta$ at $r = 1$ AU) can be related to the size and mass of the comet, and the analysis of many cometary light curves enables the mass distribution of the cometary population to be deduced. The median comet that is visible from Earth has a mass of about 10^{15} g and a nucleus of radius 1 km.

Imaging

The imaging of comets has, in the past, been mainly carried out photographically, milestones being a photograph of Donati's comet taken in 1858 by an English commercial photographer named Usherwood, and Edward Emerson Barnard's serendipitous photographic discovery of comet 1892 V on a Milky Way survey plate taken at the Lick Observatory. A fine present-day example is shown in Figure 2. Clear details can be seen

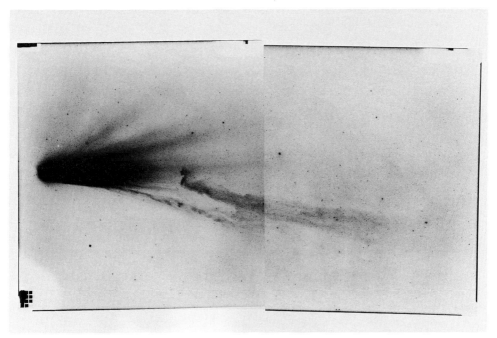

2. *Comet Halley on 10 March, 1986, 29 days after perihelion passage and 3 days before the Giotto flyby. The comet is about 0.84 AU from the Sun and 1.06 AU from Earth. The image was obtained by combining two photographic plates taken using the UK Schmidt Telescope at Siding Spring, New South Wales, Australia. Each plate covers an area of the sky of about 8° square. The smooth dust syndynes can be seen at the top of the image, as can the prominent plasma tail moving off toward the right.*

of the plasma tail and the dust syndynes (these being the curves that have their base on the nucleus on which one finds particles of a specific size that have been emitted over the previous few days). Significant variations in the cometary appearance occur from day to day. The main problem with photographic work is the logarithmic, as opposed to linear, response of the photographic emulsion when coupled with the vast range of intensities in a cometary image. The inner coma is often many powers of ten brighter than the tail, and an image like Figure 2, which is correctly exposed to bring out the structure of the tail, will completely over-expose the inner coma. Photographic imaging is still being carried out using medium sized telescopes and relatively short (a few minutes) exposures, and these pictures do provide an excellent overview of large-scale activity.

Much more computer-friendly image data can be obtained using a solid state detector at the prime focus of a large reflecting telescope. Figure 3 shows a typical example. Unfortunately the field of view is small and the monitoring of large-scale phenomena will require multiple images or the invention of much larger detectors. The accurate data given in Figure 3 is of great use when it comes to studying the physics and chemistry of the inner coma. Dust leaves the nucleus in jets, and some of the fluffy, spongy dust particles contain snow in the gaps in their porous interiors which sublimates slowly as the dust travels away from the emission region. In certain cases this snow could act as a glue and its loss could cause the particle to break up into a shower of smaller pieces. Images taken at wavelengths that are not those of the fluorescing gases show the amount of sunlight

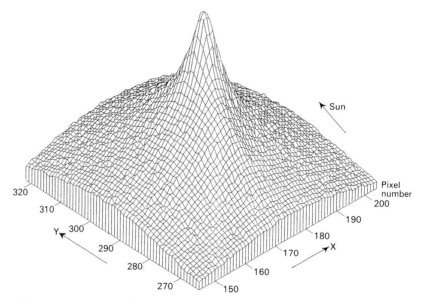

3. The inner coma of P/Halley imaged using a CCD at the prime focus of the Anglo-Australian Telescope on 11 March, 1986, 18h 14m UT. Each pixel has sides of 0.49 sec arc. The height of the 'mountain' is directly proportional to the pixel count (i.e. the DN number). The comet–Sun distance was 0.8693 AU and the Comet–Earth distance was 1.0182 AU. The x-axis is in the direction of increasing declination and the y-axis in the direction of increasing right ascension. The nucleus–Sun direction is inclined at 12.5° to the y-axis.

scattered by the dust and can be interpreted to give both the spatial density of the dust and the size distribution, as a function of distance from the nucleus. Much of the coma gas sublimates directly from the nucleus and leaves at a thermal velocity of around 0.5 km/s. These molecules are quickly broken up by photodissociation, and it is the daughter products of this process, hydrogen, OH, C_2 and C_3 that dominate the coma radiation. Cometary images are used to check the physics and chemistry of this process.

Spectroscopy

The spectroscopy of comets has been carried out ever since Giovanni B. Donati examined Comet Tempel (1864 II) and Sir William Huggins (in 1868) identified the three prominent visual spectral bands as the Swan bands of hydrocarbons. Considerable advances have been made in this discipline throughout the 1980s, and the narrow visual region between 3500 and 7500 Å has been augmented by extensive cometary observations obtained using the International Ultraviolet Explorer satellite in the 1200 to 3200 Å band and both the telescopes on the summit of Mauna Kea in Hawaii and the 36-inch reflector on the Kuiper Airborne Observatory, in the infra-red. The main spectroscopic task is to quantify the composition of the gas and dust emitted by the cometary nucleus and to convert the line strengths and continuum background in a series of spectra into a nucleus emission rate as a function of time and heliocentric distance. Compositional changes as a function of distance from the nucleus can be obtained, as can the dust-to-gas mass ratio of the material emitted by the comet. Observations of different comets and of different apparitions of the same comet are used to investigate if all comets are made of the same material and if the material in a specific comet varies as a function of depth in the nucleus. The task is not easy, and the infra-red seems to be especially difficult to interpret. The 3.28 and 3.37 micron emission features and two major absorption bands, around 3.4 and 10 microns, have been attributed to a host of organic, graphite and silaceous mixtures. The organic material might be simple carbon–hydrogen molecules such as aromatic koranine, kerogene, polyoxymethelene or polyformaldehyde compounds. Some scientists, however, have suggested that the material might be more complicated and that the spectra are indicative of polysaccharides and bacterial agents. Unfortunately different physical states of the same material as well as a range of different chemical compositions can produce similar infra-red features. Greater spectral resolution and much more work with laboratory samples are needed to make the picture clearer.

Comets from space

The space age started with Sputnik in 1957, but to date only one comet has been investigated.

Imagine that you wished to study the human race in detail but circumstances dictated that you could only have a close look at *one* human being. Whom would you choose? Would they be male or female, young or old, athletic or sedentary, beautiful or plain, ordinary or unusual? The cometary community did not have the luxury of choice, they had Comet Halley picked for them by the tax-payer in the street – a generous person who had heard of this unusual object that punctuates astronomical history every three score years and ten, and was prepared to acquiesce to the spending of the hundreds of millions of pounds needed for the interplanetary space mission to visit it. Instead of an ordinary comet, the community got the most famous comet of them all, a comet that is much brighter and larger than the normal and also, with its 76 year period and retrograde orbit, in a rather small class of intermediate comets sandwiched between the short- and long-period classes. Five space-craft were sent to Comet Halley. The only one to get within 1000 km of the nucleus and to take clear images was the European Space Agency's Giotto space-craft. It flew past Halley on 13/14 March, 1986, about four weeks after the comet's closest approach to the Sun and at a time when its heliocentric distance was 0.89 AU. The closest distance between the space-craft and the comet was 596 ± 2 km and the flyby took place at a velocity of 68.4 km/s. This was much faster than the 1 km/s of the cometary gas and the few hundred metres per second of the dust, so the space-craft saw an essentially static situation as it moved across the coma. The space-craft not only added the third dimension to cometary studies but it also improved the resolution by many decades. For the first time observers could escape from the ground-based telescope that always showed a comet 'squashed flat' against the plane of the sky. Scientific instruments could now go through the comet and measure parameters along the line of sight. A cometary nucleus one kilometre or so across, at a typical comet–Earth distance, is way below the resolution limit of a ground-based telescope. But from 596 km the nucleus becomes a tangible visible reality.

The major success of the Giotto mission was the imaging of the cometary nucleus. Prior to Giotto we had little firm idea as to the source of cometary activity. In 1835 John F. W. Herschel had seen bright jets coming from the inner coma of Comet Halley. He pictured a single solid nucleus from which jets of gas and dust were erupting. But in 1866 the dust particles responsible for the Leonid and Perseid meteor showers were found to have orbits that coincided with those of comets 1866 I and 1862 III. Overnight comets became nothing more than a vast co-orbiting bee-swarm of dust, pebbles and rocks, shepherded together by their own weak gravitation. It was not until 1951 that a strong move was made back towards the unitary model. Fred L. Whipple suggested that comets had at their core a single, solid, spinning, kilometric-sized 'dirty snowball' nucleus and that jets of gas from this nucleus, emitted from regions that did not coincide with the subsolar point, were responsible for the non-gravitational forces that led to some comets being accelerated as they passed perihelion and others being retarded. The images obtained with the camera on board Giotto (see Figure 4) showed that the nucleus was shaped rather like a smooth avocado pear having dimensions of $16 \times 8.5 \times 8.2$ km. The nucleus was a uniform dark grey, and all that was visible was dust or uniformly dirty snow. There were no white ice fields. The smoothness of the nucleus was unexpected and can be judged from the limb profile shown in Figure 4. Many discrete features such as 'craters', 'valleys' and 'mountains' can be seen. The 'crater' is no more than a 1600 m depression which is as little as 100 m deep. The 'mountain' is about 500 m high. The source of the central jet is about 50% brighter than the surrounding surface. The geometric albedo is about 4%, making the nucleus one of the darkest objects in the solar system, matched only by some carbon-rich asteroids and the dark lava flows on the Moon. This low reflectivity could be

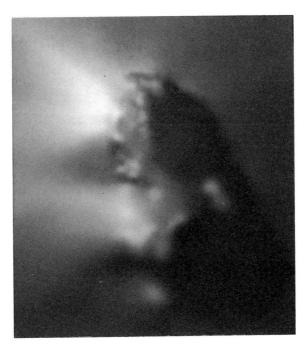

THE NUCLEUS OF COMET HALLEY

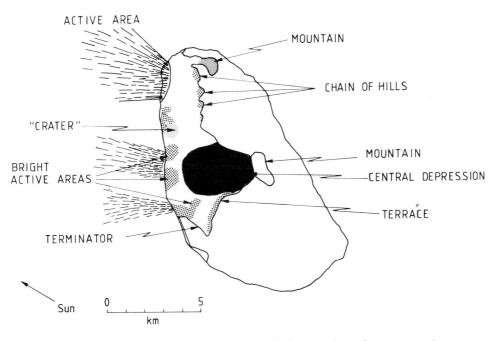

4. *This composite image of the nucleus of Comet Halley was obtained using images from the Halley Multicolour Camera on board the Giotto space-craft. North is up and the Sun is to the left and 15° below the image plane. The resolution varies from 800 m in the lower right to 100 m in the vicinity of the northern tip. The schematic drawing highlights the major features recognisable in the photograph.*

due to the porous, crumbly and spongy structure of the nucleus surface, coupled with the possibility that the dust is coated with a black tar-like organic material. Only about 10% of the total surface area of 400 km^2 was emitting dust, and this was divided up between four active areas which contained in total around thirteen faint narrow jets. It was difficult to assess whether all the surface within the area of activity emitted dust or whether the dust was ejected out of numerous small spots within that region. A gas production rate of 7×10^{29} molecules/s was estimated, equivalent to a total mass loss of 6 tonnes/s. (The mass of gas being ejected was about three times the mass of dust.) The total mass and mean density of the nucleus of Comet Halley have been estimated to be about 10^{17} g and 0.2 g/cm^3, respectively. The nucleus seems to be spinning with a period of 54 ± 1 hours, but the spin axis is precessing with a period of about 7.4 days.

It must be remembered that Comet Halley is a middle-aged comet and decay at previous apparitions has produced a meteoroid stream of mass around 3×10^{16} g. This indicates that the comet has already had about 3000 perihelion passages on its present orbit and is now only about half the size that it was when it was first captured by Jupiter. The present shape is thus a poor indicator of its initial shape, but it seems likely that mass loss has led to a general rounding off of its features.

Many data were collected during the flyby mission concerning the dust, gas and plasma surrounding the comet, but there were few surprises. Even when searching for clues as to the cometary birth place, it was found that isotopic ratios such as carbon-12 to carbon-13 and sulphur-32 to sulphur-34 were similar to those found in both terrestrial sea water and meteorites from the asteroid belt. Different parts of our Galaxy probably have different isotopic ratios, so these results point firmly to an origin of comets within the solar system. Dust mass spectrometers found that some of the 0.02–5 micron particles were made up of only carbon, hydrogen, oxygen and nitrogen. The magnesium, silicon, calcium and iron of normal rocks was missing. So comets are not just emitting dust and gas but also snowflakes and snow–dust mixtures. In the inner coma around nineteen out of twenty of the gas molecules were simply H_2O.

Many mysteries still remain. Our knowledge of the cometary nucleus is only skin deep, and we have no idea of the internal structure. The nucleus was probably formed at the dawn of the solar system by the accretion of smaller 'dirty snowballs' orbiting in the Saturn to Neptune region of the pre-planetary nebula. So it most likely contains a loosely bound collection of subnuclei, these possibly having slightly different densities and compositions. Gaps and holes abound, and the sudden breakthrough into one of these hidden caves could reveal fresh snow and result in an outburst of activity. The space-craft investigations of Comet Halley only provided a single snapshot of its activity, and we have little detailed knowledge as to how the nucleus surface varies during the many months that it spends in the vicinity of the Sun. We have also seen only one comet in some detail, and there is considerable debate as to whether Comet Halley is typical of other comets or a rather unusual member of the cometary family.

Two new cometary missions are being planned, and the hope is that they will help solve some of the questions posed earlier. The first is NASA's Comet Rendezvous Asteroid Flyby (CRAF) mission, which is due for launch around 1997. This is essentially a comet orbiter. Instead of indulging in a single 150 000 miles per hour flyby, taking a few hours to cross through the comet, the CRAF space-craft will meet a short-period comet as it comes in past the orbit of Mars and stay with that comet for ever after. Small onboard rockets will enable the space-craft to manoeuvre throughout the coma, hover above the surface of the nucleus, and finally land softly. The results should provide a detailed insight into about two years of the activity of a comet.

An even more ambitious mission is one of the four cornerstone missions in the European Space Agency's Horizon 2000 space plan. This is called Comet Nucleus Sample Return (CNSR). As the name implies, the idea is to rendezvous with a comet, land on the nucleus, excavate some of the deep-freeze cometary material and then return to Earth with the material in a pristine condition. A section of one of the basic building blocks left over from the formation of the outer planets will then be available for laboratory investigation, and quite a lot will be then known about three comets. Our ignorance concerning the details of cometary physics is well illustrated by the fact that some of the scientific space engineers planning the CNSR mission are designing rock drills whereas others are thinking in terms of ice-cream scoops.

The high resolution and multi-wavelength ability of space telescopes should enable us to integrate the detailed knowledge obtained from interplanetary cometary space-probes with a much expanded set of data about the other short-period comets and the new long-period ones.

Meteors

Visual meteors are produced in the Earth's upper atmosphere (at heights of 110 to 80 km) when dust particles of mass greater than about 0.01 g impact with velocities between 11.2 and 72 km/s. In a fraction of a second the surface of the incoming dust particle starts to boil. Molecules and atoms leaving this surface have such a high energy that they can excite and ionise the next twenty or so air molecules with which they collide. This produces a meteor that is typically 15 km long and a few metres in diameter. This relationship between the visual magnitude of the train at its brightest point and the mass and velocity of the incoming dust particle is complex, the luminous power of the train being proportional to the meteoroid's mass to the power of around 0.7 and velocity to a power between 5 and 6.

Two types of meteors are easily recognisable. The first are the shower meteors: these are associated with a specific orbit in the solar system and appear to the observer on Earth to be shooting out of a single spot in the sky, known as the radiant. At the present time the major showers are the Quadrantids, Eta Aquarids, Perseids and Geminids, and these produce, respectively, about ninety, forty five, eighty and eighty visual meteors per hour at the time of their maximum activity. The Eta Aquarids and Perseids are 'old' showers, there being, for example, twelve recorded apparitions of the Perseids between AD 36 and 1451. The Quadrantids and Geminids were, however, first recorded in 1835 and 1862, respectively. It was around this time that the two streams of meteoroids started to be swept across the Earth's orbit by the gravitational perturbations of Jupiter.

Measurements of the shower meteor flux as a function of time are made by amateur observers all over the world. The flux can be converted into a value for the spatial density

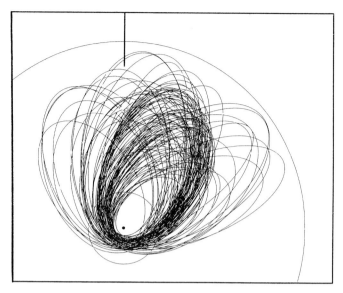

5. *The orbits of the known Southern Taurid meteoroids are shown, drawn in their mean orbital plane. The marker at the top of the diagram represents the First Point of Aries. The dot at the centre represents the Sun, and the two circles have radii of 1 AU (the semi-major axis of the Earth's orbit) and 5.2 AU (the semi-major axis of the orbit of Jupiter). These meteoroids have been produced by the decay of Comet Encke.*

of the stream. If we know how the number of meteors varies as a function of the visual magnitude, and also the orbital parameters of the stream, it is possible to calculate the total mass of meteoroid dust in a specific stream. It has been found that the Quadrantid, Eta Aquarid/Orionid, Perseid and Geminid streams have masses of 1.3×10^{15}, 3.1×10^{17}, 3.3×10^{16} and 1.6×10^{16} g, respectively. An analysis of the mass distribution of the dust in the streams indicates that about 48% of the total stream mass is made up of particles in the 0.01 to 1 g range, and increasing the range by two more decades (0.001 to 0.01 and 1 to 10 g) adds a further 25%. Only 17% of the total mass is made of particles with individual masses outside the 0.0001 to 100 g range. The visual meteor observer is thus sensitive to the majority of the stream mass. Meteor streams have been, and in many cases are still being, fed by decaying comets. The mass lost by these comets can be divided into three portions: the gas mass, the mass of the big dust particles (the ones that go into the stream), and the mass of the small dust particles (the ones that go to form the dust tails shown in Figure 2). The mass percentages that fall into each of these regions are approximately 70, 30 and 0.1. From the figures given above it is expected that the short-period comets responsible for the major meteor streams have an average present-day mass of 4.5×10^{16} g, and had an average mass just after jovian capture of 3.6×10^{17} g. Needless to say, there are many less massive comets and these produce minor meteor showers.

The evolution of a meteoroid stream is a rather messy business. The dust is given a range of velocities with respect to the cometary nucleus, and this leads to a set of orbits

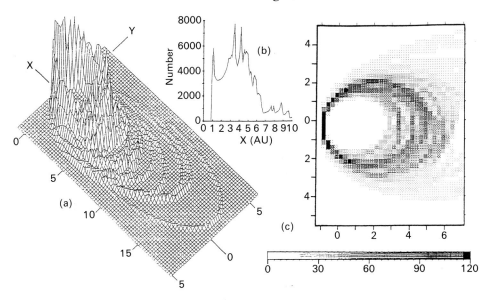

6. *The vertical direction in (a) is proportional to the spatial density of the Quadrantid meteoroids. The ground plane is the mean orbital plane of the stream. The axes are numbered in AU, and each square has sides of length 0.25 AU. The graph, (b), shows the instantaneous number of Quadrantid meteoroids, plotted as a function of distance along the X-axis (the axis that contains the mean semi-major axis of the observed Quadrantid stream). The contributions from individual orbits with large values of semi-major axis are clearly seen in the region 7–10 AU. The shaded plot, (c), shows the inner solar system region of (a). The range of shading represents the meteoroid spatial densities on a linear scale. The units are arbitrary.*

with a large range of aphelion distances. Also, the orbit of the parent comet is being perturbed and forced to precess by the gravitational influence of the planets. The effect of this is shown in Figure 5 for the meteoroids responsible for the Southern Taurid shower. The Quadrantids are slightly less perturbed. The spatial distribution of Quadrantid meteoroids is shown in Figure 6. It can be seen that the number of meteors per unit volume is greatest near the perihelion of the stream, and the high Quadrantid visual rate that is observed from Earth is due to the fact that we intersect the stream very close to its perihelion. The Quadrantid orbit has an ascending node close to the orbit of Jupiter, and this accounts for the high rate of precession and the fact that the stream will only be seen from Earth for about 300 years.

The inner solar system contains many comets and thus many meteor streams. It is only the short-period comets that we can see — the bright ones that get close to the Sun and that are actively decaying and thus losing dust. The orbits of these comets are shown in Figure 7. If we assume that each comet has produced a typical meteor stream we can then calculate the way in which the stream dust is distributed in space. This is shown in Figure 8. Notice that the Earth is not at the most favourable position as far as meteors go. The

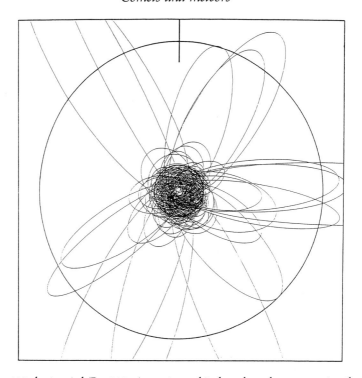

7. *The 135 short-period (P < 200 yr) cometary orbits have been drawn assuming that these comets all have zero inclination (this is a reasonable assumption for the vast majority of this group of comets, which has a median inclination to the ecliptic of 11°). The Sun is at the centre of the figure and the marker at the top indicates the direction of the First Point of Aries. The cometary orbits are orientated such that the angle, First Point of Aries–Sun–perihelion, is the sum of the longitude of the ascending node Ω and the argument of perihelion ω in the case of inclinations i between 0° and 90°. This angle is put to (Ω-ω) for 90° < i < 180°.*

The two bold circles have radii of 5.2 and 30.1 AU respectively. Moving clockwise from the First Point of Aries the cometary orbits that cross the 30.1 AU circle are those of comets P/Dubiago, P/Väisälä II, P/Olbers, P/Wilk, P/Pons–Brooks, P/de Vico, P/Brorsen–Metcalf, P/Herschel–Rigollet, P/Swift–Tuttle, P/Barnard II, P/Halley, P/Bradfield and P/Mellish.

meteoroid spatial density at the orbit of Mars is 2.2. times greater, whereas the spatial density at Jupiter is only 75% of the 1 AU value.

The jumble of orbits shown in Figure 7 indicates that the inner solar system meteoroid traffic is subject to many collisions. These lead to the breakup of stream meteoroids and the production of less massive particles on more random orbits. These fragments are hitting the Earth all the time, and can be seen as a background of sporadic meteors. Go outside on any clear dark night and you will notice the sporadic meteor flux building up to a maximum of about fifteen per hour around 0500. At that time you are on the leading side of Earth and the atmosphere is ploughing into cometary dust.

David W. Hughes

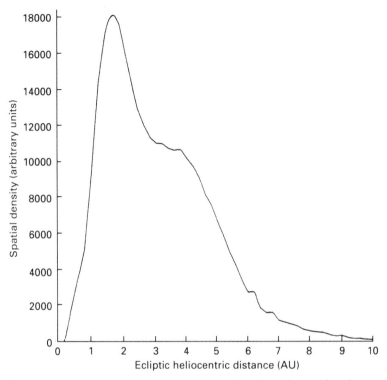

8. *The mean spatial density of stream meteoroids in the ecliptic plane of the solar system, plotted as a function of heliocentric distance.*

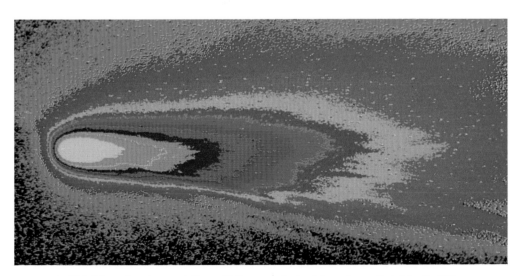

9. *This false-colour image of Comet Halley was obtained from data collected by the UK Schmidt Telescope in Australia on 4 March, 1986. The colours represent different levels of light intensity.*

91

Conclusion

Astronomers' fascination with comets stretches back to the dawn of civilisation; there is little to match the spectacular beauty of a giant comet when it, only too rarely, visits our skies. But a comet is much more than just a decaying dirty snowball — a celestial litter lout that has a host of physical, chemical and magneto-hydrodynamic interactions with the interplanetary environment of the inner solar system. A cometary nucleus is a vital and close-at-hand deep-frozen record of the early epochs of star and planetary formation, and thus provides a vital clue to one of the most exciting processes in astrophysics.

We are living in exciting times for the cometary and meteor scientist. Giotto has given us our first fleeting glimpse of a cometary nucleus, all be it from a minimum distance of 596 km. Future missions will take us much closer to the nucleus, and we will stay there for years as opposed to hours. Not only will we be able to see how a comet decays, we will also be able to witness the production of meteoroid streams. Comets represent the fundamental building blocks of the outer planets, and the cometary material has been unaltered by the ravishes of differentiation and thermal modification since the time of planetary formation. We have started to capture cometary dust; the Long Duration Exposure Facility was rescued from its Earth orbit on 12 January, 1990, by the Space Shuttle Columbia (STS-32) after nearly six years collecting dust particles from near-Earth space. In twenty years' time we hope to do much better, and I foresee freezer boxes full of cometary material being returned to Earth. Maybe then the seven questions that I posed at the beginning of this chapter will be answerable.

7

The Sun

IAIN NICOLSON

Although the Sun appears in every way to be an average star, it is of supreme interest to astronomers because it is so much nearer than any other star and therefore is the one star whose features and characteristics can be studied in great detail.

The Sun is a gaseous globe with a radius of 696 000 km (109 times the radius of the Earth), a mass of 2×10^{30} kg (some 330 000 Earth masses), and a mean density of 1400 kg/m³ (about one-quarter of the Earth's density). Its chemical composition, by mass, is about 73.5% hydrogen, 25% helium and about 1.5% heavier elements; of the heavier elements, the most abundant include carbon, oxygen, nitrogen, neon and iron. The effective surface temperature of the Sun is about 5800 K, and its luminosity – the total amount of energy radiated into space every second – is 3.86×10^{26} watts. At the Earth's distance of just under 150 million km, the quantity of solar energy falling vertically on a one square metre surface is 1368 watts, a quantity known as the solar constant.

Practically all the Sun's visible light is radiated from a thin layer known as the photosphere. Below the photosphere, solar material is completely opaque, while above the photosphere the solar atmosphere is almost completely transparent to visible light. The visible solar spectrum consists of a continuous rainbow band of colour – the continuum – together with thousands of dark absorption lines. Each chemical element produces its own characteristic pattern of absorption lines. Studies of these lines reveal

1. *Spectrum of sunlight. Dark lines occur because selected wavelengths are absorbed by chemical elements in the Sun's atmosphere. For example the Hα and Hβ lines are due to hydrogen, the D lines to sodium and the b lines to magnesium.*

not only the chemical composition of the photosphere and atmosphere, but yield a wide variety of other information, including details of temperature, pressure, gas motions and magnetic fields.

The Sun radiates electromagnetic radiation of all wavelengths from gamma rays and X-rays through to radio waves. The great majority of the Sun's radiation is emitted from the photosphere at near-ultra-violet, visible, and infra-red wavelengths. However, significant and variable amounts of radio and microwave, ultra-violet, extreme ultra-violet, X-ray and even gamma radiations originate from various levels in the solar atmosphere, usually from the chromosphere, the layer immediately above the photosphere, or the corona, the tenuous outer region of the solar atmosphere. Solar ultra-violet, X- and gamma radiation is absorbed high in the terrestrial atmosphere and cannot be studied from ground level. Information about the Sun is also provided by the streams of atomic and subatomic particles which it emits. During the past few decades, rockets, satellites and space-craft have given astronomers access to the full range of solar radiations, and the resultant images have transformed our knowledge and understanding of the Sun.

Inside the Sun

Theoretical models suggest that the central values of temperatures and pressure are, respectively, $1.5 \times 10^7 \, \text{K}$ and $1.6 \times 10^5 \, \text{kg/m}^3$ (160 times that of water). Under these extreme conditions, nuclei of hydrogen are fused together to form nuclei of helium. Each reaction, in effect, welds together four hydrogen nuclei (each consisting of a proton) to form a helium nucleus (which consists of two protons and two neutrons). In the process, two of the positively charged protons are converted into neutral neutrons by the ejection of two positrons, lightweight particles similar to electrons but with opposite (positive) charge; these positrons quickly annihilate in collisions with electrons in the solar interior.

The mass of a helium nucleus is about 0.7% less than the mass of the particles which went into its formation. This discrepancy in mass is liberated as energy in accordance with Albert Einstein's famous relationship, $E = mc^2$ whereby, if a mass m is converted to energy, the energy (E) released is equal to the product of the mass and the square of the speed of light (c). In order to maintain its luminosity, the Sun destroys about 4.4 million tonnes of matter every second. The Sun is believed to have been shining for some five billion years, but is thought to have sufficient hydrogen fuel to sustain its energy output with only very gradual changes for a further five or six billion years.

Even in the extreme conditions of the solar core, the probability of four protons

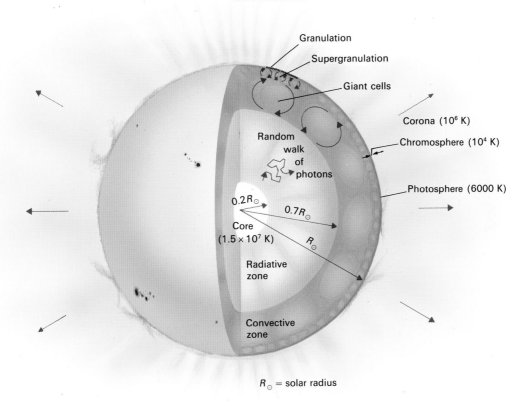

2. The structure of the Sun.

simultaneously colliding with each other with sufficient energy to overcome the mutual repulsion of like-charged particles is completely negligible. Instead, the fusion process proceeds by a series of 'two body' interactions. The dominant fusion process inside the Sun is believed to be the proton–proton reaction which in its normal mode (the PPI chain) is a three-stage process. This, together with the rarer and more complex PPII and PPIII chains, is illustrated in Figure 3.

Energy is transported outwards from the core by energetic photons, but deep inside the Sun photons travel only microscopic distances between collisions with particles of matter. At each encounter a photon either rebounds (i.e. is 'scattered') in a random direction, or is absorbed and re-emitted in a random direction, so that it follows a 'random walk' towards the surface rather than travelling radially outwards. The number of collisions involved is so great that an individual photon may take millions of years to make its way from the core to the surface; thereafter it takes only 8.3 minutes to reach the Earth. Photons lose energy in these collisions so that by the time they reach the photosphere, most of them have been converted from gamma and X-ray to visible and infra-red radiation.

Deep inside the Sun, matter is almost completely ionised; i.e. atoms have lost most or

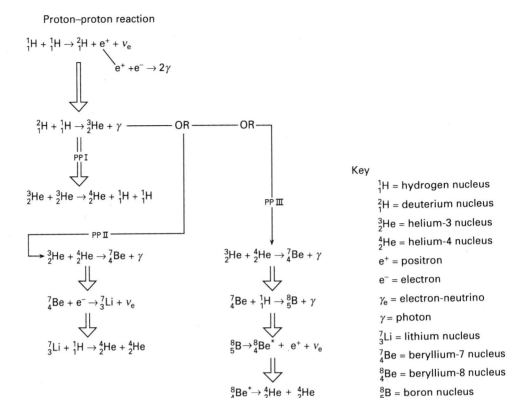

Proton–proton reaction

$$^1_1H + {}^1_1H \rightarrow {}^2_1H + e^+ + \nu_e$$

$$e^+ + e^- \rightarrow 2\gamma$$

$$^2_1H + {}^1_1H \rightarrow {}^3_2He + \gamma \quad —— OR —— OR—$$

PP I

$$^3_2He + {}^3_2He \rightarrow {}^4_2He + {}^1_1H + {}^1_1H$$

PP III

— PP II —

$$^3_2He + {}^4_2He \rightarrow {}^7_4Be + \gamma$$

$$^3_2He + {}^4_2He \rightarrow {}^7_4Be + \gamma$$

$$^7_4Be + e^- \rightarrow {}^7_3Li + \nu_e$$

$$^7_4Be + {}^1_1H \rightarrow {}^8_5B + \gamma$$

$$^7_3Li + {}^1_1H \rightarrow {}^4_2He + {}^4_2He$$

$$^8_5B \rightarrow {}^8_4Be^* + e^+ + \nu_e$$

$$^8_4Be^* \rightarrow {}^4_2He + {}^4_2He$$

Key

^1_1H = hydrogen nucleus

^2_1H = deuterium nucleus

^3_2He = helium-3 nucleus

^4_2He = helium-4 nucleus

e^+ = positron

e^- = electron

γ_e = electron-neutrino

γ = photon

^7_3Li = lithium nucleus

^7_4Be = beryllium-7 nucleus

^8_4Be = beryllium-8 nucleus

^8_5B = boron nucleus

3. *The three stage proton–proton reaction (PPI chain) which is the dominant fusion process inside the Sun, with the more complex PPII and PPIII chains.*

all of their electrons. Nearer the surface, temperatures become progressively lower, and it becomes possible for nuclei to capture electrons to form ions or neutral atoms. Matter in these states is more effective at absorbing radiation and impeding the outward flow of radiation. Energy trapped in this way causes bubbles of hot gas to rise towards the surface. As hot gas reaches the photosphere, it radiates energy to space, cools and sinks down to be reheated once more. This process of convection is the main means of completing the transfer of energy from the solar interior to the surface.

In summary, the standard model of the Sun suggests that the solar core, which extends to about 20% of the solar radius, is surrounded by a radiative zone (through which energy is transported by the random walk of photons) which extends to about 70% of the solar radius, and a convective zone which occupies the outermost 30% of the solar interior.

Although the solar interior is opaque to all forms of electromagnetic radiation, it is almost completely transparent to neutrinos. These elusive particles scarcely ever interact with ordinary matter, and readily escape directly from the solar core into interplanetary space, travelling at, or very close to, the speed of light.

It is difficult, but not impossible, to devise detectors which can capture a microscopic fraction of these neutrinos. For example, if an atom of chlorine-37 (an isotope of chlorine) interacts with a neutrino, it is changed into an atom of radioactive argon. As an individual chlorine-37 atom on average is likely to capture a solar neutrino only once in 10^{29} years, a workable detector needs to contain a very large number of chlorine atoms. Since 1965, a detector of this kind, consisting of 380 000 litres of dry cleaning fluid, carbon tetrachloride (C_2Cl_4), has been operating a mile underground (to shield it from extraneous particles) in the Homestake Mine, South Dakota. If the standard model of the solar interior were correct, it should capture about one neutrino per day. In practice, it consistently registers about one-third of the expected number of neutrinos. This discrepancy constitutes the solar neutrino problem, a problem which has been challenging solar physicists for a quarter of a century.

Does the low observed neutrino flux indicate that there is something wrong with the standard model of the Sun, or something wrong with our understanding of neutrinos?

The flux of neutrinos is very strongly dependent on the central temperature in the Sun. A central temperature 10% less than the standard figure of 15 million K would reduce the neutrino flux to the observed level. However, if the central temperature is reduced by 10% it becomes difficult, if not impossible, to produce a solar model which matches the observed values of radius, temperature and luminosity. To support a body of the Sun's radius, an additional source of pressure would be required to compensate for the reduced pressure exerted by the cooler gas (suggestions have included rapid core rotation or powerful magnetic fields in the core), and to maintain the observed luminosity at lower core temperatures would require the chemical composition of the core to be different from what is normally assumed. Perhaps the neutrino flux indicates that energy generation in the core is currently running at a reduced rate but, because radiation takes so long to diffuse to the surface, the visible luminosity has yet to respond to this reduction.

If neutrinos have tiny but finite masses, they may oscillate between three different types: electron neutrinos, muon neutrinos and tauon neutrinos. The neutrinos produced in the proton–proton reaction are of the electron neutrino type; if, in transit from the solar core to the Earth, they distribute themselves into the three different types, then the chlorine detector, which registers only electron neutrinos, will detect only about one-third of the total neutrino flux. Although this suggestion has considerable support, it cannot as yet be regarded as 'the solution' to the solar neutrino problem.

An alternative suggestion is that the solar core may contain weakly interacting massive particles, or WIMPS, which, like neutrinos, rarely interact with ordinary matter but which, unlike neutrinos, have substantial masses and would move relatively slowly inside the solar core. Collisions between WIMPS and high-energy nuclei in the central region of the core would impart energy to the WIMPS and cool the ordinary matter; subsequent collisions between WIMPS and less energetic nuclei in the outer core would impart energy to those nuclei and raise the temperature in the outer part of the core. By making the core temperature more uniform throughout, the same total amount of nuclear energy could be generated from it, but the temperature at its centre would be less; this is just what is needed to reduce the neutrino flux.

New experimental techniques should soon be able to determine whether or not WIMPS actually exist, but until that evidence is available the WIMP solution to the solar neutrino problem should be regarded as promising but 'not proven'.

A relatively new and rapidly developing means of probing the solar interior is helioseismology. Since the 1960s, astronomers have known that the solar globe is vibrating, like the skin of a drum or a ringing gong, with periods ranging from a few minutes to a few hours, but principally in the range 2.5 to 11 minutes. These oscillations show up as small periodic Doppler shifts in the wavelengths of spectral lines as localised regions of the surface rise towards and fall away from the observer.

The surface oscillations are produced by seismic, or sound, waves which propagate through the solar globe and which are probably triggered by convection. The speed of sound depends on temperature and density, both of which increase with increasing depth below the solar surface. Consequently, a sound wave moving inwards from a point on the surface is refracted and eventually curves back to meet the solar surface at another point. The sharp change in density at the surface reflects the wave back into the solar interior and, in this way, a wave can bounce around the Sun, and may interfere with itself to produce a standing wave. The greater the penetration depth of the wave, the fewer the points at which it meets the surface. By examining the thousands of oscillations and separating out modes which penetrate to different depths, solar physicists are beginning to be able to study the internal structure of the Sun in much the same way as geophysicists study the interior of the Earth. In principle, information may be gained about the variation with depth of parameters such as temperature, density, and chemical composition. Results obtained so far indicate that the convective zone has a thickness of 0.3 solar radii (considerably deeper than used to be believed) and that the initial abundance of helium in the interior was probably the same as the surface value (so apparently ruling out explanations of the neutrino problem which require different initial abundances in the interior).

By studying waves which propagate with and against the rotation of the Sun, information may also be extracted on how the Sun's rotation varies with depth. Astronomers have long recognised that the surface of the Sun exhibits differential rotation; the rotation period increases from about twenty-five days at the equator to about thirty-six days at the poles. Helioseismological results published in 1989 indicate that differential rotation extends to the base of the convective zone, but that the solar interior below this depth rotates with a uniform period of twenty-seven days.

New projects currently being developed, notably the GONG (Global Oscillation Network Group) project, offer real prospects of obtaining detailed 'images' of the solar interior and so providing stringent tests of the various theoretical models.

The surface and beyond

Optically, the photosphere reveals its partially transparent character through the phenomenon of limb darkening. At the centre of the solar disc an observer's line of sight penetrates vertically to slightly deeper, denser and hotter layers of the photosphere, while the line of sight to the edge (or 'limb') of the Sun looks at a tangent into higher, cooler and more rarefied layers. Consequently the brightness of the disc declines towards its edge.

Under really good seeing conditions, the photosphere reveals a small-scale mottled structure of bright granules separated by darker lanes. Individual granules measure 1000 to 2000 km in diameter and persist for only about ten minutes before dissolving to be replaced by new ones. Each granule represents a cell where hot gas rises to the surface, spreads out, cools and then sinks (through the surrounding lanes) back into the convection zone to be reheated.

Very narrow-band filters can isolate upward (blue-shifted) and downward (red-shifted) gas motions, and 'velocity' images of the solar surface can be constructed. Images of this kind also reveal a slower, larger-scale convective pattern, known as supergranulation, with cells some 20 000 to 30 000 km in diameter which extend deeper into the convective zone than the small-scale granules. Many solar physicists believe that there also exist giant cells, some 200 000 to 300 000 km in diameter, which extend to the base of the convective zone and slowly dredge up material from these great depths. However, the theory of solar convection is still far from complete.

The most obvious features of the photosphere are sunspots, dark patches which range in size from tiny pores, not much larger than individual granules, to complex groups covering more than a billion square kilometres. They are cooler than the surrounding photosphere and appear dark by contrast with their brilliant surroundings. Substantial sunspots have a darker central region, the umbra, where the temperature may be as low as 4000 K, surrounded by a less dark penumbra of some 5500 K; the ambient photospheric temperature is about 6000 K. Sunspots are regions of concentrated magnetic fields with strengths of up to 0.4 tesla (about 10 000 times the strength of the magnetic field at the Earth's surface); it is thought that sunspots are cool mainly because strong magnetic fields inhibit the normal convective flow of energy.

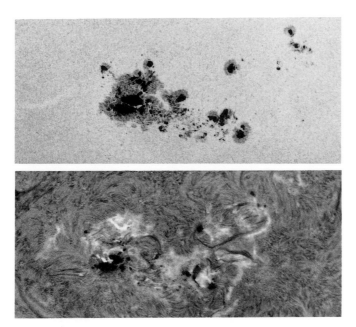

4. *These two photographs of the same group of sunspots were made by Jean Dragesco on 15 June 1989. The white light image (above) was made at 08h 42m U T, and the hydrogen alpha image (below) just over two hours later.*

Spots usually occur in pairs, or more complex groups. The pattern of magnetic field lines in a sunspot pair is similar to that associated with a bar magnet: one spot has north magnetic polarity (with outward-directed magnetic field lines) and the other has south polarity (inward-directed field lines). Groups divide into more complex regions of opposite polarity. Spot pairs and groups are the visible symptoms of the presence of bipolar magnetic regions, or active regions, on the Sun. Bipolar magnetic regions form before spots appear, and persist for some time after their disappearance. Faculae and plages are bright patches of material, concentrated and heated by the magnetic field, which accompany bipolar magnetic regions.

Solar magnetic fields are revealed by the Zeeman effect whereby a single spectral line is split into two or more components if a magnetic field is present in the region within which the line is formed. Magnetograms, images of the whole solar disc showing the distribution and polarities of magnetic regions, reveal that all the sunspot groups in the northern hemisphere have the same polarity pattern, and all the spot groups in the southern hemisphere have the opposite pattern. Thus, if the leading spot ('leading' in the sense of the Sun's rotation) of each spot pair has north magnetic polarity, the following spot in each pair will have south polarity; all the spot pairs in the other

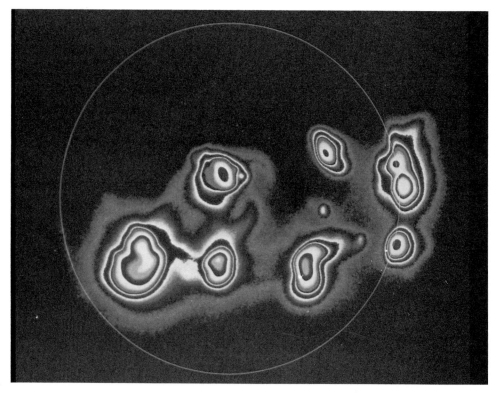

5. *This X-ray view of the solar surface reveals parallel bands of activity straddling the solar equator. Magnetic fields of opposite polarity are visible and illustrate that the leading sunspots in one hemisphere are of opposite polarity to those in the other hemisphere.*

hemispheres will have south polarity leading and north polarity following. The entire polarity pattern on the solar surface reverses every eleven years or so.

The tenuous layer which lies above the photosphere is known as the chromosphere ('colour sphere') because of the pinky-reddish hue which it briefly displays when the photosphere is obscured by the Moon during a total solar eclipse. Chromospheric gas is too tenuous to emit significant amounts of white light, but when viewed at the limb, just beyond the edge of the photosphere, displays a spectrum of bright emission lines, including the red hydrogen-alpha (Hα) line (one of several lines characteristic of the element hydrogen) which contributes strongly towards the colour of the chromosphere.

The structure of the chromosphere can be studied in monochromatic light by using a spectroheliograph or a narrow-band filter to isolate light corresponding to a particular spectral line and thereby cutting out the background glare of the photosphere. At the central wavelength of a line, the absorption is very strong and only light from relatively high altitudes in the chromosphere can escape to the observer. At wavelengths away from the centre of the line, the absorption decreases, and light is received from lower levels in the solar atmosphere. Thus images of the Sun made at the centre and away from the centre of a prominent line, such as Hα or calcium-K, reveal the structure at different

6. *The Sun in hydrogen alpha light taken at Big Bear Solar Observatory in California on 10 March, 1989. In the lower part of the disc is one of the greatest sunspot groups ever observed, it is surrounded by a large bright solar flare. A surge of dark material is being ejected at upper right. The elongated dark clouds are solar prominences, the bright patches are areas of enhanced magnetic field.*

levels in the chromosphere. Furthermore, different lines are prominent under different conditions of temperature and pressure. By selecting lines produced under different conditions, it is possible to sample atmospheric conditions over a range of altitudes and build up a vertical profile of the solar atmosphere.

The temperature in the solar atmosphere declines from about 6400 K at the base of the photosphere to 4200 K at the base of the chromosphere. Thereafter, it increases, slowly at first, then extremely rapidly in the narrow transition zone between the chromosphere and corona where the temperature surges upwards by a factor of about one hundred from about 10 000 K to about one million K.

The question of why the upper chromosphere and corona should have temperatures so much higher than the photosphere is a long-standing problem in solar physics. If energy flows outwards from the solar interior, one would expect the temperature to decline with increasing height above the photosphere. For a time, a favoured idea was that the heating process was basically mechanical, i.e. that turbulent gas motions in the photosphere and the surging motion of spicules sent shock waves into the solar atmosphere which deposited energy there and so raised the temperature. Recent studies suggest that this is unlikely to be a major process but that, instead, magnetic processes are the major heating source for the chromosphere and corona. Disturbances propagating along magnetic field lines (magnetohydrodynamic waves), heating caused by electrical currents flowing along field lines, and sudden localised releases of energy due to magnetic reconnection (discussed later in the context of flares) may all have a role to play in the elucidation of this intriguing, but as yet unresolved, problem.

The corona is a plasma, a mixture of equal numbers of positively and negatively

7. *This total eclipse of the Sun was photographed by Dan Turton from Java in June 1983. The Sun's disc has been obscured by the Moon, revealing the Sun's corona.*

charged particles consisting predominantly of protons (hydrogen nuclei) and electrons, which has temperatures ranging between one and five million kelvins. In ordinary white light, the corona has about one-millionth of the photosphere's brilliance and can only be seen during a total solar eclipse or with specialised coronagraphs which produce artificial eclipses. The white light corona extends to several solar radii, its extent and shape varying significantly over the solar cycle. The corona's optical radiation consists of a mixture of photospheric light scattered from the fast-moving electrons, and emission lines characteristic of highly ionised atoms.

Its high-temperature plasma radiates mainly at extreme ultra-violet and X-ray wavelengths. At these wavelengths, the much cooler photosphere appears dark and there is no difficulty in observing the structure of the corona against the solar disc. X-ray and extreme ultra-violet images of the Sun have transformed our knowledge and understanding of the high-temperature upper solar atmosphere. The Sun's radio emissions also emanate from the corona, and radio studies at metre wavelengths confirm the high temperatures and reveal the radio-emitting structures which are present.

The structures of the chromosphere and corona are determined by magnetic forces. Charged particles (ions, protons and electrons) can flow along field lines but cannot flow at right angles to them. Where the density and pressure of a plasma are low, its motion is controlled by the magnetic field, and a moving field carries the plasma with it; conversely, the flow of a denser plasma carries the magnetic field lines with it. Under these circumstances, the magnetic field is said to be 'frozen' to the plasma. The interplay between fields and plasmas determines almost all aspects of structure and activity in the solar atmosphere.

Viewed at the limb in monochromatic light, the chromosphere is seen to contain large numbers of flame-like columns of gas known as spicules. Typically 10 000 km high and 1000 km thick, they rise and fall along local magnetic lines of force and persist, on average, for five or ten minutes. Monochromatic images also reveal spicules as dark blade-like absorption features against the solar disc. Clumps of spicules delineate a cellular pattern known as the chromospheric network which is linked to the underlying photospheric supergranules.

Long dark filaments, sometimes stretching for as much as 200 000 km, are also seen in monochromatic images. These represent denser clouds of gas, suspended in the solar atmosphere and usually aligned along the neutral line separating areas of opposite magnetic polarity. When seen beyond the solar limb, the filaments appear as luminous prominences. These clouds of hydrogen, suspended by magnetic field structures, have temperatures of 7000 to 10 000 K, and are typically one hundred times denser than the surrounding coronal material. There are two basic classes of prominence – quiescent and active. Quiescent prominences hang suspended in the corona for weeks or months on end with little change in their overall structure. Active prominences show rapid and dramatic changes which can catapult tongues or loops of gas to heights of 300 000 km or more and, in some cases, eject material into interplanetary space.

X-ray and extreme ultra-violet images of the Sun reveal that the corona consists of bright regions of concentrated high-temperature plasma confined by magnetic loops, quiet regions of less intense emission, and dark coronal holes, low-density regions where

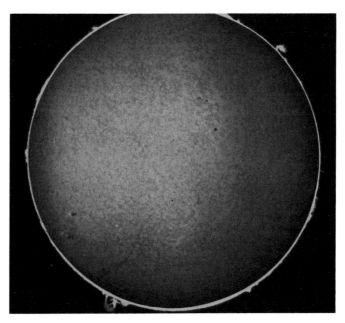

8. *Solar granulation, sunspots, filaments and prominences are all visible on this picture of the Sun taken by H. J. P. Arnold on 18 July, 1989. Two hydrogen alpha images − one exposed for disc detail and the other to record prominences − were combined to produce this picture.*

magnetic field lines drift out into interplanetary space, so allowing solar plasma to escape into the interplanetary medium. The general flow of plasma (mainly protons and electrons) from the Sun constitutes the solar wind, which 'blows' past the Earth at speeds ranging from 200 to 900 km/s. The Sun loses about one million tonnes of mass per second in this way. The solar wind and interplanetary field have been investigated by many space-craft; hopefully its three-dimensional structure will be revealed by the 'Ulysses' space-craft which is currently following a route, via Jupiter, that will enable it to fly over both solar poles.

The most violent of solar phenomena are flares − explosive releases of energy in which up to 10^{25} joules (equivalent to detonating ten billion one-megatonne nuclear devices) can be liberated within a few thousand seconds. Flares emit radiation over practically the whole electromagnetic spectrum from hard (short wavelength) X-rays, or in some cases gamma rays, to radio wavelengths; the bulk of their radiation is emitted in the X-ray and extreme ultra-violet wavebands. They eject streams of energetic atomic particles including electrons − which may be accelerated to half the speed of light − and nuclei (mainly protons), and expel bulk clouds of plasma through the corona. As the high-speed streams and clouds plough through the corona they stimulate the emission of microwave and radio radiations over a wide range of frequencies.

Flares are believed to be caused by the sudden release of magnetic energy, stored in the twisted magnetic fields of complex sunspot groups, by magnetic reconnection. This occurs when oppositely directed field lines come into contact and sever to form two loops, or a loop connected to the solar surface below the point of contact, and an open U-shaped field structure above. In the process, part of the magnetic field is 'annihilated' and converted into other forms of energy – thermal (heat) and kinetic (particle streams and bulk motion). Localised heating at the flare site produces much of the X-ray and extreme ultra-violet radiation, particles accelerated down the loop produce hard X-rays and Hα emission as they plough into the low corona and chromosphere, and electrons accelerated upwards along field lines stimulate bursts of radio emission in the corona. Like an elastic string, a magnetic field line has tension in it. When it is severed and reconnected, it catapults bulk plasma outwards, possibly to escape into interplanetary space.

All forms of solar activity, sunspots, plages, prominences and flares, together with the structure of the chromosphere and corona, show cyclic variations with a period of about eleven years, but the most obvious indicator of fluctuating solar activity is the sunspot cycle. The number of sunspots reaches a maximum roughly once every eleven years, and at the intervening minima few if any spots may be seen for weeks on end. The level of activity in successive cycles can differ significantly, and there appears to be some evidence for longer-term modulation of activity on time scales of eighty years or more. Between 1645 and 1715, a period known as the Maunder minimum, solar activity seems virtually to have ceased, and there is some evidence for similar lapses in the cycle in the more

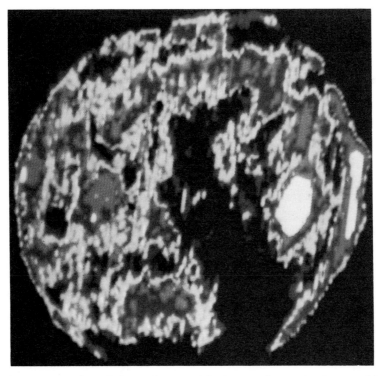

9. *This computer-enhanced ultra-violet view of the Sun reveals a spectacular coronal hole. It extends from the northern polar region to below the solar equator.*

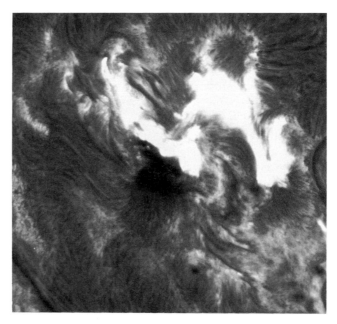

10. *Two ribbon flares in the solar chromosphere photographed in hydrogen alpha light on 17 June, 1989.*

remote past. Sunspots usually occur in two latitude bands; the first spots of a new cycle normally appear at about 30–40 degrees either side of the solar equator, and as the cycle progresses the bands of activity progress towards the equator. The polarity pattern of sunspot groups reverses every eleven years, so that in terms of repeating the basic underlying magnetic pattern, the Sun has a twenty-two-year cycle.

The solar magnetic field, which controls this cycle, is believed to be generated and sustained by a dynamo process driven by circulating currents in the convection zone, but solar physicists as yet do not really understand what produces the cyclic winding up and relaxation of solar activity. One model, first devised by Horace W. Babcock in 1961 and subsequently developed by Robert Leighton, accounts for the broad features of the cycle as follows. If at some stage of the cycle magnetic lines of force enter the Sun in high northern latitudes, pass along meridians under the surface and emerge in high southern latitudes, and if those lines are considered to be frozen to the solar material, then differential rotation will stretch and distort them, eventually wrapping them many times round the solar disc, and amplifying the field where the lines are bunched together. When the field in a bundle of lines, or 'flux tube', becomes sufficiently strong, the flux tube will float to the surface, producing a pair of spots. Where the field lines emerge, a spot is formed with outward-directed field lines (north polarity), and where they re-enter the solar surface a spot will form with south polarity.

The winding action of differential rotation drags the magnetically active bands towards the equator as the cycle progresses. As individual spot groups decay, the polarity of the following spot preferentially diffuses towards the polar region in its hemisphere. The accumulation of polarity eventually changes the overall polarity of the Sun and a new eleven-year cycle begins. Although attractive in outline, many details of the theory are uncertain or incomplete.

An alternative theory, proposed in 1987 by Herschel B. Snodgrass and Peter R. Wilson, suggests that deep-seated convection by giant cells of 200 000 km scale, is the driving force which produces the cycle. If giant cells originate at the poles and gradually roll down towards the equator over some twenty-two years, they will mesh together like gears so that adjacent cells rotate in opposite directions. Where two flows come together, field lines are squeezed into bunches which become sufficiently concentrated by the time the cells have rolled down towards 40–30 degrees of latitude, to become buoyant and form sunspot groups. The subsequent motion of the cells drags the sunspot zone towards the equator where the cells dissipate.

The solar activity cycle has direct influence on the Earth. Fluctuations in the solar wind compress the Earth's magnetosphere to produce disturbances in the surface magnetic field, known as magnetic storms, and resultant surges in power lines and telephone cables. Changes in atmospheric vorticity (storminess) and thunderstorm activity can be linked to fluctuations in the solar particle output. At times of high solar activity, when flares are more frequent, the solar output of ultra-violet, X-ray and particle radiations can be dramatically enhanced; this increases the ionisation in the terrestrial atmosphere with resultant, and sometimes dramatic, effects on radio communications. Violent particle outbursts from the Sun disturb the magnetosphere and catapult charged particles down the Earth's magnetic field lines into the upper atmosphere to create the shimmering aurorae. The enhanced heating of the upper atmosphere (the thermosphere) by solar ultra-violet and X-radiation causes the upper atmosphere to expand and thereby exert an increased drag on orbiting satellites; this process led, for example, to the orbital decay and destruction of the Skylab space station and the Solar Maximum Mission (SMM) satellite.

Precise measurements of the solar constant made from satellites such as Nimbus 7 and SMM show that the solar energy output fluctuates on time scales of days or weeks by up to 0.2% due, in the main, to changes in the number of dark sunspots or bright faculae on the disc. However, measurements over the past decade have revealed a more general variation in solar luminosity which follows the solar cycle; the Sun is more luminous by about 0.08% at solar maximum than at solar minimum. It would appear that when solar activity is at a maximum, the total energy output from bright faculae, plages, enhanced ultra-violet and other short-wave radiations more than compensates for the light loss due to dark sunspots. Recent observations have also shown that much of the variation seems to be due to the presence of bands of marginally increased surface temperature (about 1 K warmer than the neighbouring photosphere) which migrate towards the equator with the sunspots.

In terms of its magnetic field and its X-ray, ultra-violet and radio output, the Sun can be regarded as a variable star. Although its cyclic variation in total luminosity is slightly less than 0.1%, this variation lends credence to the historical correlation between very depressed solar activity during the 'Maunder minimum' from 1645 to 1715, and unusually cold European winters in the middle of that period – the so-called 'little ice age'. Whether such fluctuations are also linked to minor variations in the solar radius remains 'not proven'.

Solar physics

Solar physics has made dramatic progress over the past two centuries. In the early nineteenth century, astronomers had no real physical understanding of the composition and structure of the Sun, or of its energy source. The Sun was widely held to be a dark body, perhaps even inhabited, surrounded by a luminous layer; sunspots were thought to be holes in the luminous envelope. Prominences and the corona had been seen at eclipses but their nature remained a mystery and no one knew whether the corona belonged to the Sun, the Moon, or the Earth's atmosphere.

Many of these puzzles were unravelled with the advent of the spectroscope. William Wollaston first noted dark lines in the solar spectrum in 1802, and by 1814 Joseph von Fraunhofer had observed no less than 574 such lines. The true significance of these lines was not revealed until Gustav Kirchhoff and Robert Bunsen established the laws of spectroscopy in 1859 and thereby provided the key to establishing the chemical composition of the Sun. Spectroscopic observations in 1868 and 1869 showed that prominences and the corona comprised hot gas and were certainly features of the Sun. The coronal emission lines did not seem to match any known element, and for some seventy years many astronomers believed they represented a new element, 'coronium'. Finally, in 1940, Bengt Edlen showed that the lines were due to highly ionised elements, notably iron, and demonstrated, therefore, that the corona must have a temperature in excess of one million K.

The magnetic fields of sunspots were first detected in 1908. The solar magnetograph pioneered in 1952 by Harold and Horace Babcock enabled astronomers to study solar fields in detail, and to begin to discover the extent to which magnetic fields dominate all aspects of solar activity. Radio emissions from the Sun were first identified in 1942, the solar ultra-violet spectrum was first recorded by a rocket-borne instrument in 1946, and solar X-rays were first detected by a rocket flown in 1949. Since the early 1960s, satellites, space-craft and manned missions have vastly broadened and deepened our database of observational material.

The idea that plasma from the Sun might be responsible for terrestrial magnetic storms and aurorae was discussed as early as 1900 by Sir Oliver Lodge, and gained ground in the 1930s. In 1951, Ludwig Biermann invoked this mechanism to explain the behaviour of the ion tails of comets, and in 1958 Eugene Parker showed theoretically that a 'solar wind' must be flowing out of the corona. Firm confirmation of this idea came from interplanetary space-craft within a few years.

Nineteenth-century ideas on the source of the Sun's energy ranged from the infall of meteoritic material to the idea, proposed in 1854 by Hermann von Helmholtz, that slow gravitational contraction would be able to power the Sun for some twenty million years. Geological evidence showed this time scale to be far too short. By the 1920s there had been great advances in the theory of stellar structure and the idea was gaining ground that nuclear reactions involving the transmutation of elements was the key to solar energy.

By 1938 Hans Bethe and Charles Critchfield had established in detail how the fusion of hydrogen to helium would work in practice in the solar interior.

Direct investigations of the solar interior through studies of neutrinos and oscillations began to become possible from the 1960s onwards, but these studies are in their infancy and many problems remain. Solar structure, energy generation and transport, atmospheric heating, the mechanisms for flares, prominences and transients, the three-dimensional structure of the solar wind and interplanetary field, the interaction of these entities with the magnetosphere, and solar variability, are all pressing issues with which observers and theoreticians will continue to grapple in the 1990s. By improving our detailed understanding of the Sun, we may hope better to understand the chromospheres, coronae, spots, flares, variability and evolution of stars in general.

Solar physics is in an exciting phase. While advances in theory and observation eventually will solve many problems, undoubtedly they will throw up many more to ensure that solar physicists will have plenty to puzzle over for a long time to come.

8

Stellar evolution*

JACQUELINE MITTON

Imagine, if you can, what the universe might have been like if there were no such things as stars. Suppose that the rules by which matter and energy behave and interact were such that stable stars could not exist. This mental exercise might appear somewhat pointless since stars obviously do exist in abundance, but it serves to emphasise the importance of stars in shaping our perception of the nature of our universe. Stars do a great deal more than make the night sky look pretty. They generate most of the luminous energy in today's universe and, as Sir Arthur Eddington put it in 1920, 'In the stars matter has its preliminary brewing to prepare the greater variety of elements which are needed for a world of life'. Later, theoretical astrophysicists showed that processes occurring in stars are indeed responsible for the creation of many of the chemical elements.

The Earth's environment is dominated by our nearest star, the Sun. Our Sun holds unparalleled importance for us, and yet within the universe of stars there is nothing particularly special about it. It is a splendid example of an ordinary star. Nevertheless, it has a special fascination for astronomers, simply because of its proximity. By studying the Sun, astronomers have been able to unlock the more closely-guarded secrets of stars in general. So, as a first step to understanding the stars, it is helpful to explore the similarities and differences between the Sun and stellar populations at large.

* Adapted from *Invitation to Astronomy* by Simon and Jacqueline Mitton, Blackwell, 1986.

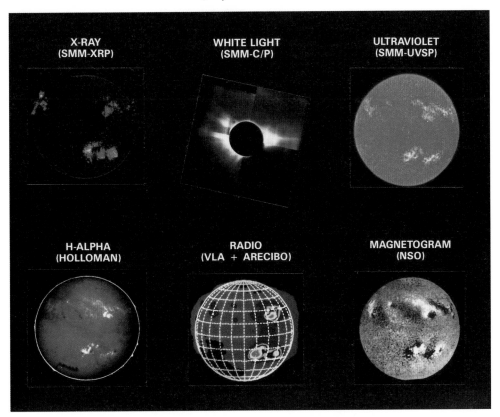

1. *The Sun in six bands of the electromagnetic spectrum, imaged 2 and 3 September, 1988. The images were produced as part of the International Solar Month campaign to study the Sun in many wavebands simultaneously, from Earth and space. The three images at the top were produced by three instruments on NASA's Solar Maximum Mission satellite. The lower three images are ground-based views of the Sun.*

Before the discovery of nuclear energy, the source of the Sun's energy presented an insurmountable problem for astronomers. Geologists were able to demonstrate that the Earth is at least four and a half billion years old, so some process was obviously at work that could keep the Sun shining over even longer time scales.

Eddington was an outstanding figure in the elucidation of this problem. His theoretical work between 1916 and 1925 demonstrated convincingly that the Sun and the stars are spheres of gas with a source of energy at the centre. By 1920 he had pinpointed nuclear processes as the likely energy source and had even suggested that the actual process might involve the transformation of hydrogen into helium with the release of energy. Referring to Lord Rutherford's atom-splitting experiments at Cambridge, he said, 'What is possible in the Cavendish Laboratory may not be too difficult in the Sun'!

By 1936, Robert d'Escort Atkinson had worked out the details of the nuclear processes by which helium and heavier elements can be synthesized from hydrogen in the Sun's interior. Atkinson was assured of an ample supply of fuel because it had been recently demonstrated that hydrogen is by far and away the most abundant element in the Sun. Some five million tonnes of matter are totally annihilated every second within the Sun, releasing energy at a rate in accordance with Einstein's famous equation, $E = mc^2$. In this equation E represents energy, m is the mass of material destroyed, and c is the velocity of light. Even at this prodigious rate, the Sun is able to keep going even in its present state for a further five billion years. In due course, however, the Sun will inescapably consume the entire supply of hydrogen fuel in its core. The question is, 'What happens then?'

Let us look at the life cycle of a star, the sequence of events from a star's creation through its life to eventual death: the process of 'stellar evolution'. (In this context, evolution is used to mean the sequence of changes taking place to a single object, whereas, in connection with living organisms, it usually means changes that are perceptible only as one generation succeeds another.)

Stars are created when some stimulus causes part of the interstellar medium to collapse. A pressure wave or turbulence in the medium might be enough. Once the process has started, the gravitational pull that each particle of matter has on the other ensures that the protostar continues to condense until other forces take over. Stars are essentially giant gas balls, so it is important to understand the rules of physics that gases follow in order to work out what goes on inside a star.

In simple terms, when a gas contracts it gets hotter and when it expands it gets cooler. So, as the initial gas ball falls in on itself, the temperature at the middle goes up. Once it reaches a certain critical temperature – about four million degrees – the hydrogen atoms are able to crash into each other so hard that they stick together to form helium. Astronomers usually call this thermonuclear fusion reaction 'hydrogen burning', although it is not chemical combustion in the sense in which a coal fire burns.

Once the nuclear reactions start, the star has a long-term internal source of energy. The heat generated by the reactions flows outwards towards the surface. At the same time, the gas pressure halts the gravitational collapse, and the star settles down into a comfortable condition of equilibrium. In this steady state, the heat energy flowing outwards is balanced by that generated in the core. Any tendency of the gas to fall inwards under gravity is exactly counteracted by the gas's need to expand outwards because it is hot. In the Sun's case this stable state will last about ten billion years in all, about half of which time has already passed.

Ultimately, the hydrogen in the core will be used up and the fusion reactions will stop. When the heat supply is turned off, the gas in the core will start to fall in on itself again. This will have the effect of generating heat as gravitational energy is liberated. When the temperature reaches a hundred million degrees a new reaction can start: the temperature and pressure will be so great that the nuclei of helium atoms will start to fuse together to form carbon and oxygen. The Sun will have a new, but temporary, lease of life.

Although the core of the Sun will collapse, the upset in the equilibrium will have the opposite effect on its outer layers: they will expand outwards, swallowing up the inner planets, including Earth, in the process. The expanding gas will get cooler and, as the temperature drops, its yellow colour will fade to red. The Sun will become a 'red giant'

112

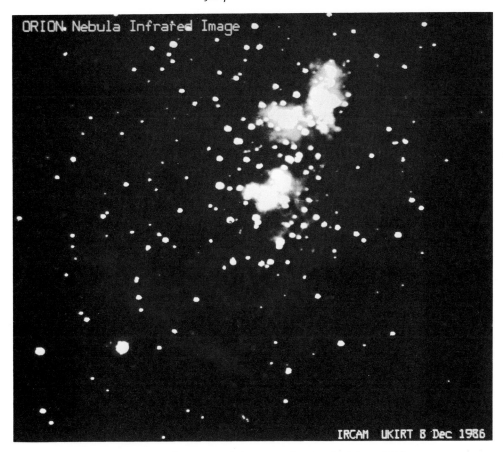

ORION Nebula Infrared Image

IRCAM UKIRT 8 Dec 1986

2. This infra-red image of the Orion Nebula was made using the Infra-red Telescope on Mauna Kea in Hawaii. In this image we see through the dust clouds which obscure our view in other wavelengths, revealing many more of the young stars within the nebula.

star. All this change will take place in a period of time that is relatively short compared with the hydrogen burning phase: only about a hundred million years.

Although the Sun is a typical star in some sense, all stars are by no means identical to the Sun. To what extent are stars fundamentally different from one another, and how much of the difference is due to the fact that they may be at different stages in their lives? It turns out that the age and evolutionary status of a star is very important in determining its colour and luminosity. Equally important is the star's mass. The more massive a star is, the more powerful its internal energy source and the hotter its outer layers get. Conversely, a relatively lightweight star will be at the cold end of the stellar temperature scale. The lowest mass limit for a star to be able to function at all is about one-tenth that of the Sun. Ironically, perhaps, the brightest stars use energy so fast that their resources are rapidly exhausted, even though they are large. Small, feeble stars have lifetimes thousands of times longer than their more massive counterparts.

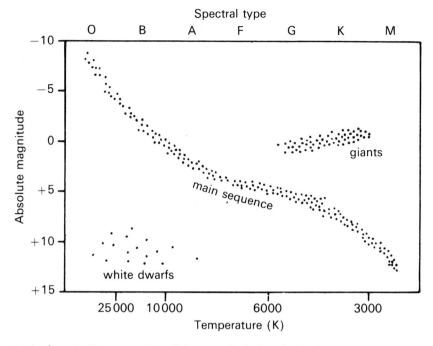

3. A schematic Hertzsprung–Russell diagram which plots the absolute magnitude of a star against its temperature. The majority of stars lie in well-defined positions on the diagram. The largest group forms the main sequence where, in general, the hotter the star, the more luminous it is. Two other well-defined groups are the cool and luminous red giants, and the very hot but much less luminous white dwarfs.

In 1913, Henry Norris Russell first described a graphical presentation relating the luminosities and colours (or spectral classes, which amounts to the same thing) of the stars. Such plots became known as 'Hertzsprung–Russell diagrams', in recognition of the fact that Ejnar Hertzsprung had in fact plotted some similar diagrams at around the same time. It was Russell, however, who realised the importance of presenting stellar data this way as a help to understanding the evolution of the stars.

In a Hertzsprung–Russell (or just HR) diagram, absolute luminosity is plotted against colour or its equivalent, such as spectral class or temperature. Each star is thus represented by a single point on the diagram. The plot takes on its true significance when it is made for certain groups of stars. These might be, for example, the stars belonging to a star cluster.

Russell's original diagram included points for all stars on which he had data. In order to work out the absolute magnitude of a star, its distance needs to be known. At that time, the stellar distances available to Russell were limited and many were very inaccurate. Nevertheless, he demonstrated clearly that the majority of stars fell in a broad band running from the upper left to the lower right of his diagram. This band has come to be known as the 'main sequence'. Above and to the right of the main sequence lay another area in which most of the remaining data points occurred. These represented stars that are relatively cool, but far more luminous than the main sequence of stars. In fact, these are the red giants. Russell deduced correctly that the difference in luminosity between the giants and the 'dwarfs' on the main sequence was due to a difference in size and density, rather than a difference in mass. However, he was mistaken in his view that the red giants were the youngest stars, still in the process of gravitational collapse. As we have seen already, they are in fact old stars.

At one time, astronomers believed that the main sequence represented an evolutionary process: it was supposed that stars moved along the main sequence as they aged. This is now known to be totally wrong. The main sequence reflects the fact that stars have a range of masses. When a star settles into its hydrogen burning phase, the more massive it is, the brighter and hotter it will be. Its position on the main sequence depends primarily on its mass. Stars burning hydrogen, like the Sun is now, are often called 'main-sequence stars' for that reason.

The notion of tracing on an HR diagram the changes in luminosity and colour that a star undergoes in the course of its lifetime has proved immensely useful to astronomers. Such a line is called an evolutionary track. The tracks are different for stars of different masses. They are also very complex, doubling back on themselves and crossing the main sequence. They are, of course, deduced from calculations since it is impossible to monitor changes in an individual star. However, the visual presentation of data in the form of HR diagrams for stars belonging to clusters provides important supporting evidence for the theorists.

The significance of a star cluster in this context is twofold: all the stars can be assumed to have the same age and to be at the same distance from us. The first factor means that the effects of the different rates of evolution experienced by stars of different mass show up; the second means that accurate distance measurements are not needed to compare the stars directly with each other. As long as they are at the same distance, apparent brightnesses are a measure of the true intrinsic luminosities.

Astronomers recognise two distinct types of star families, known as clusters: open clusters and globular clusters. Open clusters are usually loose, irregular groups containing about one hundred or so stars. Globular clusters, as their name implies, are densely packed spherical clusters containing many thousands of members. Globular clusters date back to the earliest era of star formation in our Galaxy, whereas open clusters contain much younger stars. This age difference becomes apparent when typical HR diagrams for members of the two cluster types are compared.

To understand how the diagrams can be interpreted, consider for a moment what happens to the members of a star cluster over a period of time. The masses of the stars cover a range so, at the start, they should form a nice diagonal main sequence on an HR diagram. The important thing now is that the rate at which evolution proceeds depends on the mass of the star. The most massive stars turn into red giants fastest.

When a star becomes a red giant, its data point on the HR diagram moves off the main sequence towards the upper right area. The effect of this is that the main sequence of a star cluster a few billion years old is no longer straight, but curls off towards the right at the top. This is because the most massive members have evolved off the main sequence.

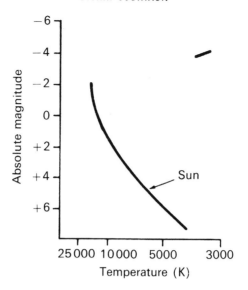

4. *A schematic Hertzsprung–Russell diagram for a young, open star cluster. Most of the stars are in the main sequence region. Some of the more massive stars have evolved into red giants, as indicated at top right. The position where the Sun would fall on this diagram is marked by the arrow.*

By measuring how great the curling is, and how much of the straight main sequence remains, astronomers work out how old the cluster is. The shorter the piece of main sequence left, the older the cluster. This is one way in which we know that globular clusters are very old, and that some open clusters are very youthful indeed.

Let us now return to the fate of the Sun; the red giant phase is only a brief respite in an inexorable trail to obscurity. Helium as a fuel soon runs out in the core and further gravitational collapse sets in. Hydrogen burning can start again in a shell around the burnt-out core. Eventually, the residual helium and hydrogen may be burnt in concentric shells that gradually move outwards as the fuel is consumed and internal adjustments to temperature and pressure occur. Ultimately, however, both fuels will become spent and the final collapse sets in.

During the changes and adjustments that occur as a star attempts to eke out its remaining nuclear fuel, instabilities develop that cause the outer layers to pulsate and perhaps to be blown off the star completely. The outer parts of a pulsating star physically bounce backwards and forwards like an oscillating spring. This causes the brightness of the star to vary in a regular way. Astronomers have recognised several different categories of variable stars whose variability is best explained by the pulsation mechanism. Historically, the most important group is probably the Cepheid variables, named after their prototype, Delta Cephei.

When a relatively low-mass star like the Sun reaches the end of the red giant phase in its life, it expels its outer envelope of tenuous gas into space, creating a glowing sphere

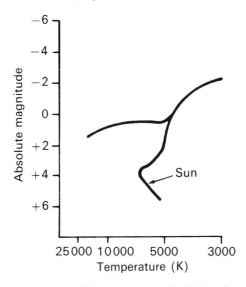

5. *A schematic Hertzsprung–Russell diagram for an old globular cluster. Most of the stars in this cluster have evolved into giants; a very few remain on the main sequence. The position where the Sun would fall on this diagram is marked by an arrow.*

around it, which gradually expands until it disperses into space. Numerous examples are known of evolved stars surrounded by spheres of nebulosity. They are known by the curious name of 'planetary nebulas'. They have nothing whatever to do with planets except that William Herschel, who invented the term, thought that their images resembled the discs of planets in the field of view of his telescope. Most planetary nebulas are roughly circular, or ring-shaped, though not all are completely symmetrical. The gas in the nebula glows because it is excited by the radiation from the star in the middle.

Gradually the central star becomes more and more compressed. There is no energy source to counteract this tendency until a phenomenally high density is reached at which a totally new physical effect comes into play. At densities of around a million times those of the ordinary matter we are familiar with, the electrons in the material act together to create a pressure that arises from their quantum mechanical properties. In effect, there is a minimum volume into which the electrons can be squeezed. This behaviour is called degeneracy.

When the Sun has collapsed into a sphere about the size of the Earth, degeneracy pressure due to the electrons will halt the process. The Sun will then be a white dwarf. On the HR diagram, white dwarfs fall well below the main sequence in the bottom left. One such star, a member of the Sigma Eridani System, was plotted by Russell on his original diagram of 1913. Numerous white dwarfs have now been identified, even though they are intrinsically very faint. Two of the brightest stars in the night sky, Sirius and Procyon, both have faint, white dwarf companions forming binary systems with them.

Many stars belong to binary systems, and the question of what happens to the system

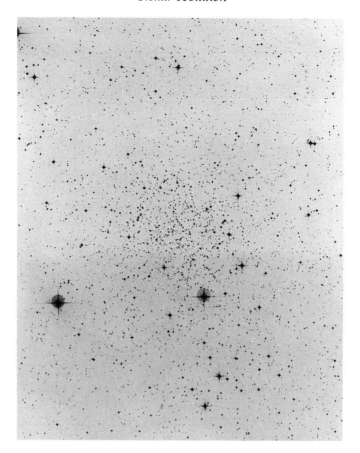

6. *The open star cluster Melotte 66. This cluster of a few hundred stars was formed between 5 and 10 billion years ago and is one of the oldest open clusters in the galaxy.*

as a whole as they evolve is interesting and important. If one member is more massive and so evolves more quickly, when it expands in its red giant phase some of its material may stream over to a close-enough companion. The spectra of some binary systems show clear evidence that interactions of this kind are going on.

When a star like the Sun becomes a white dwarf, it gradually cools down and fades away; it is a dead star. Calculations show, however, that the greatest mass that a white dwarf can have is just under one and a half times the mass of the Sun. We already know that there are many stars much more massive than that, even taking into account the fact that some material is lost to space during the course of evolution. What, then, is the fate of a massive star?

Let us go back to the stage where the hydrogen has been used up in the core. In the case of a star like the Sun, helium can take part in new reactions, but once this is used up the star is simply not large enough to get any further reactions going. Inside more massive stars, a whole sequence of reactions can take place. As each fuel is successively used up,

7. *This globular star cluster, M13, in the constellation of Hercules, is the brightest in the northern skies. The photograph was taken by Akira Fujii.*

8. *The Helix Nebula is the nearest planetary nebula to us. This Anglo-Australian Telescope image shows the spherical shell of glowing gas ejected from the central star.*

gravitational collapse raises the temperature until the next round of reactions starts. Each of these reactions creates heavier and heavier elements in turn. The ultimate effect is to create a star with a layered structure, often likened to an onion. At the centre there is an iron-rich core; round this in spherical shells are layers of silicon, oxygen, carbon, helium and hydrogen. Once the core has turned to iron, no further fusion reactions can take place like the ones that have gone on before. Deprived of an energy source, the burned-out core implodes. The sudden release of energy causes a gigantic explosion that literally blows the outer layers of the star away. A tiny neutron star may be all that remains. Such a stellar explosion is called a supernova. All stars over about one and a half solar masses end their lives in this spectacular way.

9

Variable stars

JOHN ISLES

Stars generally remain steady in brightness for most of their lives. Our own Sun, for example, is believed to have shone for almost five billion years with only slight changes in its light output, and yet it has in prospect several further billion years of near-constancy before it will begin to swell into a red giant. Apart from such slow evolutionary changes, there are certain stages in a star's life when it becomes variable in brightness on a shorter time scale (from a matter of decades down to a fraction of a second). Close binary stars can also interact in various ways that give rise to changes in their combined light output. The study of the stars that change in brightness – the variable stars – can give us insights into the structure of stars, and into how they are formed, live and die. Several types of variable star are also important 'standard candles' that help us to determine the scale of the universe.

Measuring the stars

Our Sun is the only star whose surface features can be resolved in any detail. With the aid of special techniques such as interferometry, the apparent angular diameters of some stars have been measured, but most stars can, from the observer's viewpoint, be considered as point sources of light. Consequently their surface brightness, or light emitted per unit of the star's area, cannot be measured directly.

121

The starting point is therefore to measure the amount of light received per unit area by a detector on the Earth's surface, or on a satellite in orbit about the Earth. The light received is commonly expressed on a scale of 'magnitudes', in which numerically larger values denote progressively fainter stars. These 'apparent magnitudes', as seen from the Earth, depend on the star's distance, since the apparent brightness of a given light source is inversely proportional to the square of its distance. If the distance of a star is known, the apparent magnitude can be converted to 'absolute magnitude', which is the magnitude the star would have at a standard distance of 10 parsecs (one parsec is about thirty-one million million kilometres.) A star's luminosity (its rate of energy output) is related to its absolute magnitude. Alternatively, we can express the star's light output as so many times that of our Sun; this is the star's relative luminosity.

Measurements are made by comparison with standard stars whose light is believed not to vary. The light detector may be the human eye, a photographic emulsion, or an electronic device such as the photoelectric photometer or the charge-coupled-device (CCD). The light in various ranges of wavelength may be measured separately, and the U, B, V (ultra-violet, blue, visual) system is very widely used.

Estimation by eye is still the most popular method among amateur astronomers, but it is not very accurate; the uncertainty of an observation is usually at least 0.1 magnitude, or 10% in intensity. As there are so many understudied variable stars, such observations are nevertheless very useful, especially when the range of variation is several magnitudes. Photoelectric measurements can be accurate to better than 0.01 magnitude (1%). In 1989, the most active observer of variable stars in the United Kingdom was not a human being but a computer-controlled photoelectric telescope constructed by an amateur astronomer, Jack Ells of Bexley, Kent.

As we increase the number of wavelength bands measured, we look in closer detail at the distribution of light in the star's spectrum. An extension of this principle is spectroscopy, in which the light is spread out by a prism or diffraction grating into a band in which the light is displayed in order of wavelength through all the colours of the rainbow. This spectrum can then be studied in great detail. Other techniques of observation can be used to gain further information about variable stars, but it is change in the overall light output that defines the class.

The standard catalogue of variable stars is maintained by a team of astronomers at the Sternberg State Astronomical Institute in Moscow, at present headed by Nikolai Samus. More than 30 000 variables are now listed, and thousands more stars have been suspected to vary. Most of the known variables are stars passing through a transient stage in which cyclic pulsations or eruptions occur, or are objects that do not radiate their light equally in all directions, so that their apparent brightness varies as they rotate. Several dozen types of variable stars are recognised, which are grouped into six classes: eruptive, pulsating, rotating, cataclysmic, eclipsing and X-ray. This classification does not always bring together objects that are closely related, so we shall discuss the various types in a different order here.

In this brief account we can mention only a few of the more important types, some of which can be placed in a rough sequence according to the point at which they may occur in a star's lifetime; but it must be understood that a variable star of one type does not

1. *The Orion Nebula (M42) is a site of recent star formation containing many young Orion variables. This photograph was taken by Ron Arbour using a 40-cm reflector.*

usually evolve into a variable of another type. For most of its life a typical star will not vary its level of brightness noticeably, and no star can pass through all the stages we shall mention.

New-born stars

Stars form by the collapse of clouds of interstellar matter, principally hydrogen, and young stars are often found in association with the diffuse gaseous nebulae from which they were born. Many of these stars are irregularly variable, and are known as Orion variables, because the Orion Nebula includes many such objects. The first to be found was T Orionis, by G. P. Bond in 1863. Modern catalogues list about 700 variable stars in the Orion Nebula, and they are also found in several other regions where stars have recently formed. In extreme cases these stars may change by a magnitude or more in a few hours or days, although most vary more slowly. Some show cyclic variations as the star rotates, or occasional fades like those of Algol-type eclipsing binaries, or flares like the UV Ceti stars.

An important subtype of the Orion variables is the T Tauri stars, which are distinguished by intense emission lines of neutral iron in their spectra. About half a dozen T Tauri stars have been seen to undergo prolonged outbursts, with a rise of up to six magnitudes (that is, increasing in brightness 250 times) to a maximum that lasts several decades; these are the FU Orionis stars or 'fuors'.

The variations of the Orion variables are apparently due in part to processes similar to those seen on the Sun (spots, faculae and flares), but with greater violence, and compounded with the effects of variable obscuration by clouds of dust moving near the star.

Mature stars

Once a star has settled down to become a normal 'main sequence' object or dwarf star, it seldom shows dramatic variations. At this stage it obtains its energy by nuclear fusion in its core, converting hydrogen to helium. Some main-sequence stars vary as they rotate, because their surface brightness is not uniform.

The Sun is a slightly variable star of this type. Measurements of its light output by satellites have revealed variations of up to 0.2% (0.002 magnitude) on a time scale of days or weeks, due to the appearance and disappearance of sunspots and faculae, and their passage onto and off the visible hemisphere as the Sun rotates. In addition there is a variation of 0.08% (0.0008 magnitude) correlated with the sunspot cycle. Surprisingly, the Sun is brightest at the phase of sunspot maximum.

Red dwarf stars continue to show flare activity similar to that seen in some Orion variables; these are the UV Ceti stars. They may brighten by up to six magnitudes in a few seconds, and then fade over several minutes to their original level. The flares are thought to be due to disturbances in their magnetic fields, as in the case of solar flares.

Some hot main-sequence stars, such as Gamma Cassiopeiae, are spinning rapidly, probably close to the point of rotational breakup, and from time to time they eject rings of gas around their equators, or complete shells of gas. These episodes may be accompanied by a fade or a rise in the light output; the variations can amount to as much as 1.5

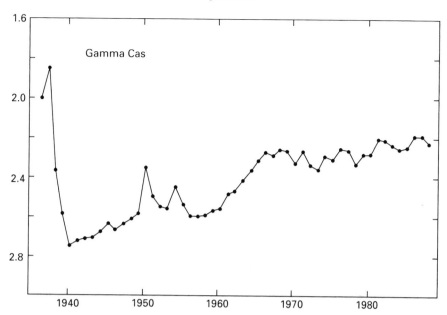

2. *The light curve of the shell star Gamma Cassiopeiae, from annual means of visual estimates by members of the British Astronomical Association and Junior Astronomical Society. There was a bright maximum in 1937, followed by a fade to minimum in 1940. Since then the brightness has risen irregularly to magnitude 2.2.*

magnitudes. Gamma Cassiopeiae itself rose to magnitude 1.6 in 1937, and faded to 3 in 1940. Since then it has climbed slowly to magnitude 2.2, but it has shown little change in the last ten years. Another well-known Gamma Cassiopeiae type star is Pleione, or BU Tauri, in the Pleiades.

Very massive stars are also unstable. These include P Cygni, Eta Carinae and S Doradus, which is visible in a good pair of binoculars, even though it is some 50 000 parsecs distant in a neighbouring galaxy, the Large Magellanic Cloud. The rate at which a star consumes its hydrogen increases greatly with its mass, and the prodigal S Doradus stars will have exhausted their supply after at most some tens of thousands of years. They are surrounded by expanding shells of gas that they have ejected. These stars radiate light at a level close to what is called the Eddington limit. If a star radiated more than this limit, the radiation pressure would blow the star apart. The instability of these very massive stars is due to the extremely delicate balance between radiation pressure and gravitation. The massive Wolf–Rayet stars can also have non-stable mass outflow and may show slight light variations.

Stars in old age

Eventually, a main-sequence star exhausts the available hydrogen in its core and begins hydrogen burning in a shell surrounding the core. At this stage the outer layers of the star expand and the surface cools, so that it becomes a red giant. The subsequent evolution of the star can be quite complicated. If it is massive enough, further nuclear reactions can begin, in which helium is converted to carbon, and then carbon into heavier elements. If the star's route is plotted in the Hertzsprung–Russell (HR) diagram that relates temperatures and absolute magnitudes, it is found to move at first upwards and to the right as it evolves away from the main sequence. Later it may move back and forth several times across the HR diagram.

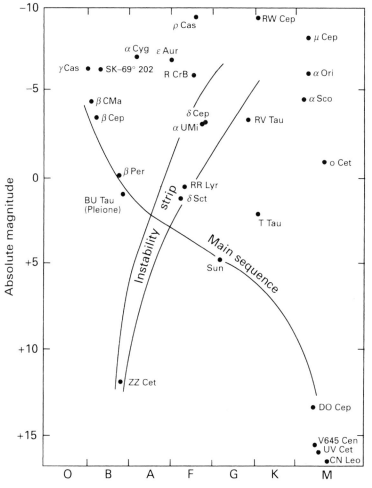

3. *A Hertzsprung–Russell diagram showing selected variable stars plotted according to their absolute magnitude and spectral type. The instability strip is a region in which many pulsating variables are found.*

There is a band in the HR diagram, known as the 'instability strip', that contains many pulsating variable stars. The best known of these are the Cepheids, named after Delta Cephei, a supergiant star whose periodic changes in a cycle of 5.37 days were first noticed by John Goodricke in 1784. For these stars, at least, the cause of the variations is understood, though many details remain to be worked out.

Energy is produced near the star's centre at a steady rate, but its flow up to the surface becomes pulsed when it passes through a layer of helium at some depth below the surface. The helium atoms alternately absorb energy, each releasing one electron and thus becoming ionised, and then release energy when they recombine to form neutral atoms of helium. If the helium layer is at the right depth below the surface to drive pulsations with the natural frequency of oscillation of the star's outer layers, this behaviour is self-perpetuating and the star becomes a pulsating variable. If the layer is at the wrong depth, such pulsations cannot build up.

This model explains the existence of the instability strip. The properties of a star, including its pulsational stability or instability, as well as its temperature and luminosity which define its position in the HR diagram, are essentially determined by only a few parameters – its mass, age and initial chemical composition. Thus it is not surprising to find that pulsating variables are found only in certain regions of the HR diagram.

Standard candles

It is also possible to use this understanding to predict theoretically (though not so far with very good agreement with observations) the relationship between a star's period and luminosity. As we look higher up the instability strip, we find stars of higher luminosity pulsating with progressively longer periods. For example, Delta Cephei with a period of 5.37 days has a visual absolute magnitude of -5, and emits about 10 000 times as much light as our Sun; l Carinae, with a period of 35.5 days, is probably about twice as luminous as Delta Cephei.

The period–luminosity (p–l) relation makes the Cepheids particularly useful as distance indicators. Even the nearest Cepheids are too remote for their distances to be measured directly by determining their parallaxes, but it is possible to estimate the mean distance for a sample of Cepheids in our own Galaxy from an analysis of their apparent angular velocities across the sky, or 'proper motions'. Their luminosities can then be calculated and used to calibrate the p–l relation. We can then use the relation to tell the luminosity of any other Cepheid whose period is known – even a Cepheid in another galaxy. As we also know its apparent magnitude, we can find the distance of that galaxy using the inverse-square law.

Several other types of pulsating variable stars can also be used as distance indicators, including the RR Lyrae stars and Mira variables. Although these are all luminous objects, if we use Earth-based telescopes we can still detect them only in the galaxies of the Local Group. Less direct methods must be used to estimate the distances of more remote galaxies. The Hubble Space Telescope, launched into Earth orbit in 1990, offers the prospect of extending the use of Cepheids and other variables as 'standard candles' out to more distant stellar systems.

Things are complicated by the fact that Cepheid variables are found among both the young stars of Population I, which forms the disc of our Galaxy, and older stars of Population II, which forms the halo and hub of the Galaxy. Population I Cepheids, such as Delta Cephei itself, contain a greater proportion of 'metals' (elements other than hydrogen and helium) and so have different properties from Population II Cepheids, such as W Virginis. Moreover, the Population I Cepheids may be crossing the instability strip

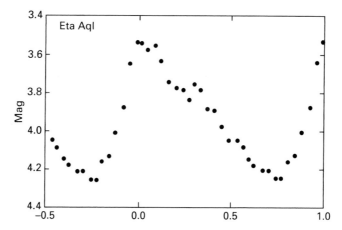

4. *The mean light curve for the Cepheid variable Eta Aquilae, period 7.18 days. Each point is the mean of eleven visual observations in 1987 and 1988 by John Isles. Note the hump on the descending branch of the curve.*

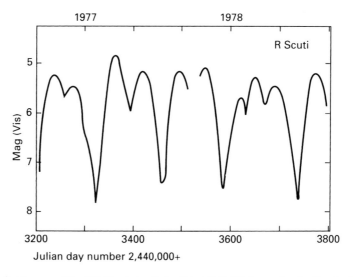

5. *The light curve of the RV Tauri star R Scuti in 1977–78, from visual estimates by members of the American Association of Variable Star Observers. The average interval between deep minima is 146 days, and there is usually a secondary minimum midway between deep ones.*

for the first time, whereas Population II Cepheids may already have been variable stars of this type several times before, as their advanced evolution has carried them back and forth in the HR diagram. The two subtypes differ in luminosity by about one magnitude (a factor of 2.5) for periods of about one day, and by two magnitudes or more (a factor of six or more) for periods longer than ten days, Population I being the brighter.

Other types of pulsating variable are also found in in the instability strip, including the subdwarf SX Phoenicis stars; the short-period Delta Scuti stars, on or just above the main sequence; the giant RR Lyrae stars or cluster variables; the RV Tauri stars, which have deep and shallow minima in rough alternation; and the semiregular variables of intermediate spectral type.

Red giants

Pulsating variables are not confined to the instability strip, however. An important group of red giants is sometimes collectively called the long-period variables. This contains the

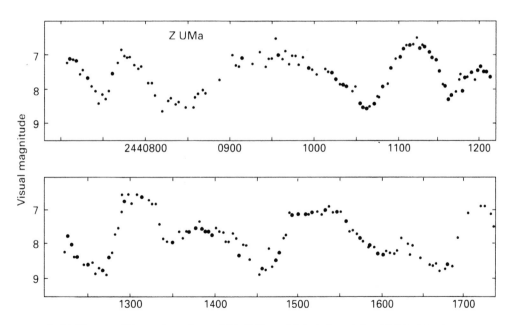

6. Light curve of the semi-regular variable Z Ursae Majoris from May 1970 to March 1973, from 4-day means of visual observations by members of three European groups: the Association Française des Observateurs des Etoiles Variables, the Berliner Arbeitsgemeinschaft für Veränderliche Sterne, and the Werkgroep Veranderlichke Sterren (based in the Netherlands). Large points are the mean of three or more estimates. There is a main period of 196 days, and usually there are two maxima and two minima in each cycle.

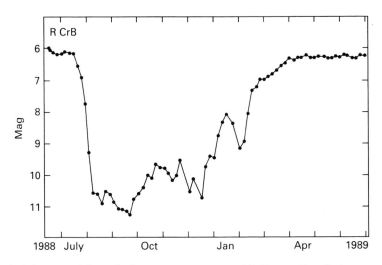

7. *The light curve of the multiple extinction minimum of R Coronae Borealis in 1988–89. The points are 5-day means of visual estimates reported to the British Astronomical Association. There were three main minima, probably due to the ejection of three sooty clouds that obscured up to 99% of the star's light.*

Mira stars, varying in a period of about one year by up to a factor of 20 000 in visual brightness (though only by a factor of two or so in total light, since at minimum the emission is mainly in the infra-red). The group also includes the red-giant semiregular and irregular variables, which may be Mira stars that have temporarily left off their major pulsations. The mechanism operating in this group of variables probably resembles that which we have seen in the Cepheids, except that it is a layer of hydrogen that drives the oscillations, rather than one of helium.

Other pulsating variables include the Beta Cephei stars, Alpha Cygni stars and PV Telescopii stars. In certain types of pulsating variables, including the white dwarf ZZ Ceti stars and at least some members of the Beta Cephei and Delta Scuti classes, it appears that the pulsations are non-radial, so that the shape of the star periodically deviates from spherical.

Near the top of the instability strip we also find the R Coronae Borealis stars. These are carbon-rich supergiants that occasionally and unpredictably fade sharply, over a few weeks, losing sometimes more than 99% of their light, before beginning a gradual recovery over several months. R Coronae Borealis itself is normally a sixth-magnitude star just visible to the naked eye, but at a deep minimum it can fall to magnitude 15, and then requires a large telescope to be detected.

The minima of R Coronae Borealis type stars are due to obscuration by a cloud of carbon-rich dust, and it is thought that one of these clouds may be ejected in a random direction every few weeks, at a certain stage in a pulsation cycle when the atmosphere 'pops'. Only if this cloud comes in our direction do we see a deep minimum. So far, the connection between the onset times of extinction minima and the phase of pulsational oscillations has not been convincingly demonstrated, and more observations are needed. The R Coronae Borealis type stars are usually regarded as eruptive variables, but if the connection of the fades with pulsations is real it may be more logical to class them with the pulsating stars.

Death throes

The final stages of stellar evolution are also accompanied by light variations. Massive stars explode as supernovae of Type II, when their cores collapse to form a neutron star. The neutron star may be observable as a pulsar, two of which have so far been detected as optical variable stars: the Crab pulsar, produced in the supernova of 1054, and the Vela pulsar, which varies in a period of 0.089 between blue magnitudes 23.2 and 25.2, making it one of the faintest variable stars known. Supernovae and pulsars are described in other chapters of this book.

In less massive stars, where the mass of the core does not exceed about 1.5 times the Sun's mass, the outer layers are ejected as a planetary nebula and the core settles down to a prolonged stage of senility as a white dwarf. The magic figure of 1.5 solar masses is called the Chandrasekhar limit; it is the maximum mass that can be supported by the pressure of the material in a white dwarf star. Above this limit, the star is likely to collapse in a supernova explosion to form a neutron star or a black hole. Even if a star survives to the white dwarf stage, its variations may not be at an end, as many white dwarfs are pulsating variables of the ZZ Ceti type.

Double variables

The types of variables we have so far considered are found among single stars, or in the members of binary systems in which the components are too far apart to have had much effect on each other's evolution. In close binaries, especially those that can exchange matter with one another, the bizarre interactions that can occur give rise to a bewildering variety of further types of variable star.

The easiest class of variable double stars to understand is the eclipsing binaries, in which one component periodically passes in front of the other, as seen from the Earth. These are usually divided into three types. In stars like Algol (Beta Persei), the variations due to eclipses occupy only a small part of the oribital period. Algol's period is 2.87 days, of which just ten hours is taken up by the primary eclipse, when the component of greater surface brightness is partially obscured by its companion. At this phase the magnitude of Algol falls from 2.1 to 3.4; there is also a secondary minimum when the companion is eclipsed by the primary star, but this is only 0.05 magnitude deep.

Many Algol-type eclipsing binaries also show what is called (incorrectly) the reflection effect: the side of one star that is bathed by its partner's radiation is at a higher temperature and appears brighter, so that even outside eclipses there is a slight change in the observed magnitude of the system as it rotates.

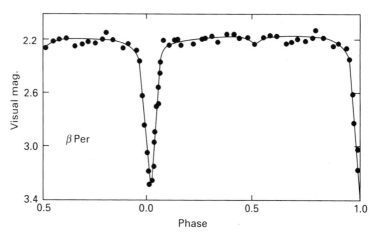

8. *Mean light curve for the eclipsing binary Algol. Each point is the mean of ten visual observations in 1987 and 1988 by John Isles. There is a shallow secondary eclipse midway between primary minima.*

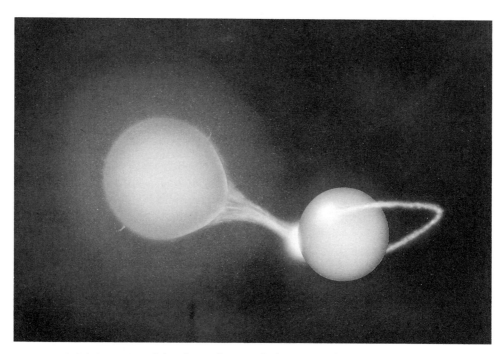

9. *Artist's impression of the eclipsing binary, Algol. Gas from the fainter component falls towards the primary and most of it lands at a 'hot-spot' on its surface. A portion of the gas misses this point and swings round the hot star, landing at a secondary hot spot on its other side. The structure of Algol, and of the interacting binaries shown in the next two illustrations, is inferred from observations; the component stars are much too close to be resolved in any telescope.*

10. The eclipsing binary Beta Lyrae as it might appear when viewed close up, from the direction of Earth, outside eclipse (above) and at primary minimum (below). Gas from the primary component falls towards the secondary and forms a dense toroidal 'accretion disc' around it.

In Beta Lyrae stars, the brighter component is ellipsoidal in shape (or, rather, egg-shaped), so that its apparent magnitude also varies outside eclipses. The W Ursae Majoris stars are contact binaries: the two components share a common envelope so that to a nearby observer the system would look something like a dumb-bell. W Ursae Majoris stars have constantly changing brightness, with light curves resembling those of Beta Lyrae stars, but with minima of nearly equal depth.

We also encounter 'reflection variables' and 'ellipsoidal variables' that resemble, respectively, the Algol and Beta Lyrae stars, but whose orbital inclination is such that eclipses are not seen from the Earth. An important subgroup of the Algol type is the RS Canum Venaticorum stars, with intense chromospheric activity and large areas of starspots on one component.

Erupting binaries

The variations in the combined light of eclipsing binaries are mainly geometric rather than due to any intrinsic change. There are also binaries with components that interact physically, giving rise to some of the most spectacular types of variable star. Most of

*11. Artist's impression of the cataclysmic variable star OY Carinae. Gas from the red
dwarf component falls towards its white dwarf companion and forms a flat accretion disc
around it. There is a hot-spot where the stream collides with the disc.*

these belong to the class of cataclysmic variables, which show eruptions at intervals
ranging from a few days up to, probably, many centuries. In most of these objects, a cool
star, usually somewhere near the main sequence, is losing mass in the direction of a hot
star, which is usually a white dwarf. The mass usually does not land directly on the surface
of the white dwarf, as it has too much angular momentum; instead, it forms a bright
'accretion disc' surrounding the dwarf. In some, the white dwarf has a strong magnetic
field which causes the mass to flow along the field lines and land in an 'accretion column'
at the star's magnetic pole.

The novae are cataclysmic variables in which the hydrogen-rich material, accumulated
from the white dwarf's partner, periodically undergoes thermonuclear detonation, being
converted to helium with a massive liberation of energy. A fast nova, such as V1500
Cygni which rose to magnitude 2 in 1975, can rise by as much as nineteen magnitudes
in only a few hours, and fades to minimum in the course of several months. Slow novae,
such as HR Delphini of 1967, may take several weeks to reach peak brightness, and the
decline can take several years.

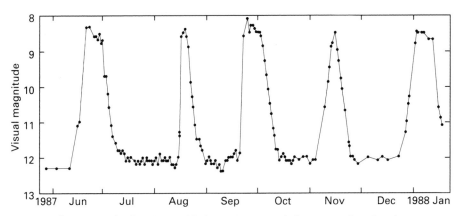

12. *Light curve of the dwarf nova SS Cygni, from visual observations by John Isles. Five outbursts occurred during the eight months of observation.*

In the dwarf novae, the outbursts are the pulsed release, in the form of light and heat, of gravitational energy by material in the accretion disc, falling to the surface of the white dwarf. The best-observed dwarf nova is SS Cygni, which rises from magnitude 12 to 8 about every fifty days. Thanks to the assiduous monitoring of this star by amateur observers, none of its outbursts have been missed since it was discovered in 1896.

Novae and dwarf novae are probably not distinct; several ex-novae, such as GK Persei which erupted in 1901, have shown minor outbursts like those of dwarf novae, while the known dwarf novae may all be potential future novae.

The 'nova-like variables' are mostly objects resembling ex-novae, in which no outbursts have so far been seen; although several objects that resemble novae at some other stage in their development are also included in this type.

Related to the cataclysmic variables are the symbiotic stars, whose cool component is a red giant, and the X-ray binaries, in which the compact component may be a neutron star or even a black hole. Only a few dozen X-ray binaries have so far been identified as optically variable stars. Among interacting binary variable stars we must also count the supernovae of Type I, which are thought to be white dwarf components that undergo collapse to a neutron star. As they all release the same amount of energy, they have the same peak luminosity, and so can be used to estimate the distances of their parent galaxies in the same way as for Cepheid variables, but to much greater distances.

Unravelling the behaviour of interacting binary stars is an important area of modern astronomical research in which amateur astronomers play a vital role. The collected results of the members of organisations such as the American Association of Variable Star Observers, the British Astronomical Association and the Royal Astronomical Society of New Zealand provide continuous light curves for many cataclysmic variables that are an important aid to the interpretation of these systems. Many novae are first detected by amateurs, who are also usually the first to report the rare outbursts of the dwarf novae that have long cycles, and the bright, long 'supermaxima' of SU Ursae Majoris systems – a subclass of the dwarf novae – during which critical observations can be made with professional equipment.

Observing stellar evolution

In the long run, every star is variable. The changes in luminosity that occur as a star evolves are usually too slow to show up in available astronomical records, but there are a few possible exceptions. Pleione, in the Pleiades, has already been mentioned as a slightly variable object of Gamma Cassiopeiae type. If it has faded over the centuries, it may account for the fact that the Pleiades cluster is traditionally known as the Seven Sisters, although there are now only six stars clearly visible to the unaided eye. Greek legend indeed records that one of the Pleiades faded to invisibility, but the original count of seven may have been determined merely by the magical associations of that number, rather than by careful observation.

There are, however, some better-documented discrepancies between the current magnitudes of the naked-eye stars and those given in older lists, particularly Ptolemy's catalogue, which was based on observations made by Hipparchus in the second century BC. None of these suspected 'secular variables' has been fully confirmed, as we have accurate magnitude records only for the last century or so; but there is evidence that the bright supergiants have brightened by up to a magnitude since Hipparchus recorded them. This is consistent with modern theories on rates of stellar evolution.

As the baseline of our observations of stellar magnitudes extends, we shall eventually be able to document stellar evolution directly. This will take many centuries. In the meantime, the study of variable stars at particular crises during their lives will continue to yield valuable information about the processes governing the lives and the deaths of stars.

Variable-star studies come of age

The study of variable stars is in itself evolving as new methods of observation become available to astronomers. The fact that some stars vary in brightness was first established four centuries ago. Several stars that we would now call novae and supernovae had been recorded by early observers, but these were generally believed to be atmospheric phenomena. The scientific study of variable stars began with Tycho Brahe's positional measurements of the supernova of 1572, which showed that it had no detectable parallax and so must be a remote object. The first stars to be recognised as showing periodic changes were Mira Ceti, by J.K. Holwarda in 1638; Chi Cygni, by Gottfried Kirch in 1686; and Algol, by John Goodricke in 1782. By 1850 about forty variables had been discovered from visual observations.

The application of photography to the stars multiplied the number of discoveries enormously. Regular photographic sky patrols were instituted in several countries, following the lead of the Harvard Observatory in 1890. At about the same time, the spectroscope offered the prospect of determining the composition of the stars, and gave rise to the new science of astrophysics.

It is only during the last forty years that astronomers have succeeded in putting together the main pieces of the jigsaw of stellar evolution, leading to a clearer understanding of many classes of variable stars. In the last thirty years, rocket-borne experiments and satellites in Earth orbit have enabled observations to be made at wavelengths to which our atmosphere is opaque, and many new classes of object have been recognised. Until 1987, stars other than our own Sun could be studied only by analysing their electromagnetic radiation. This was changed on 23 February 1987, at 07.36 Universal Time, when a burst of neutrinos from Supernova 1987A in the Large Magellanic Cloud was recorded by detectors in Japan, the USA and the USSR.

No less important than the new observational techniques is the availability of computing power to analyse the data and to explore the consequences of theoretical models of variable stars. Although the main details of many classes of variable stars seem to be becoming much clearer, the pace of change is such that in coming years we can be sure of seeing many more exciting developments in the study of variable stars.

10

Supernovae

PAUL MURDIN

The word *nova* literally means a new star and a *supernova* (plural supernovae or, in the USA, supernovas) is a very bright new star. 'New' means newly discovered – a bright star appears prominently where there was no noticeable star before. But there had been a star at that place – an old star, in fact, not a new one at all. Supernovae appear suddenly, rising in brightness quickly, over a day or two, to their most prominent; it is this rapid and large increase of brightness which makes them seem as if newly born. But over a month or two they gradually fade below visibility.

Supernovae are stellar cataclysms, stellar explosions which, in some cases, represent the end of a star's normal life and the creation from it of a compact stellar cinder. In other cases, supernovae represent the total destruction of a star. The explosions are the most powerful which occur in stars; in a supernova the amount of energy released is equivalent to the total annihilation of the mass of tens of thousands of Earths, several per cent of the mass of the Sun. This energy is released in about one second; the luminosity of a supernova for that second rivals the luminosity of all the light from all the other stars in the universe.

There is a supernova somewhere in the universe every second or so; most are far away and literally out of sight. But some supernovae are so close and so bright that, on some rare occasions, they force themselves to be noticed.

In 1572 the Danish astronomer Tycho Brahe was driving in his carriage when he saw

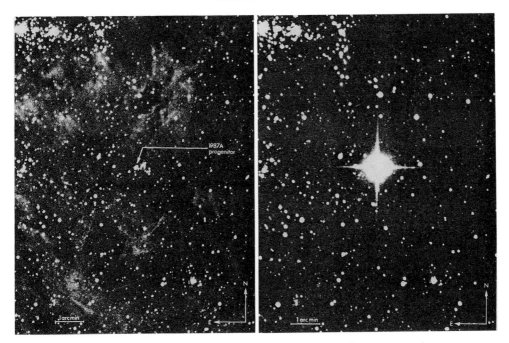

1. *Sanduleak -69 202 lies at the arrow tip in the 'before' picture; the star now no longer exists, the image of its exploding death as the Large Magellanic Cloud supernova being caught in the 'after' picture.*

some peasants gazing up to the night sky, engaged in animated discussion. He recorded that there, in the familiar constellation of Cassiopeia, was an extra star, easily seen since it disturbed the well-known W-shape of the constellation. It is known as Tycho's supernova.

In 1604 a supernova appeared in the vicinity of an astrologically significant conjunction between Mars and Jupiter, which was being observed by all astrologers everywhere, who were bound to notice the new star of planetary brightness which was suddenly present nearby. It became known as Kepler's supernova, since he collated and published the observations about it.

These supernovae together became very famous, the subject of common discussion by philosophers and poets, by educated people everywhere – by all who were interested in astronomy (the group of people who would today be called amateur astronomers). These two supernovae showed clearly that changes could take place amongst the stars, and that the stars were not eternal.

Both these supernovae occurred before the invention of the telescope. In the last millennium only half a dozen supernovae have been so bright as to be seen with the unaided eye – since the invention of the telescope, only one. It occurred in 1987: it was the first supernova of 1987, SN 1987A in the Large Magellanic Cloud. The previous naked eye supernova was the one which occurred nearly 400 years before, in 1604.

From the long times recorded between bright supernovae, the natural and correct deduction is that supernovae are rare: there is a supernova in our own Galaxy, or its nearest neighbours, and therefore potentially bright enough to be readily seen with the naked eye, only once every one hundred years on average. Most of the 600 supernovae which have been recorded have occurred in galaxies outside our own, their light diminished by distance, their outstanding power reduced by the awesome dimensions of intergalactic space from the amazed adjectives of somebody's diary to the dry calculations of an astronomer's notebook.

Discovery of supernovae

Supernovae in external galaxies are often discovered accidentally. An astronomer slews his telescope to a galaxy which he wishes to study. There in the eyepiece or on the TV screen is the image of a star which is not in the photograph from a previous epoch which the astronomer has brought to the observing run. He has stumbled upon a supernova.

If, for a year, a single telescope looks at a hundred galaxies (chosen at random) once a month (so that if a supernova occurs in one of the galaxies it is likely to be discovered before it has faded), then the effect is the same as monitoring a single galaxy for a hundred years, and spotting the, on average, one supernova in it. Perhaps there are a hundred telescopes in the world of large enough size to discover supernovae, and one hundred supernovae should be discovered per year – the average is about twenty-five per year, so perhaps most large telescopes look at less than one hundred galaxies per month.

This seems to be the way most supernovae are discovered. But it is unsystematic and random. Dedicated searches have the best chances of finding supernovae, and give the most reliable results on how frequently they occur.

The first systematic search for supernovae was organised by Fritz Zwicky, the astronomer who first recognised supernovae for what they were and who coined the word for them. His vision inspired the Palomar Observatory Supernova Search, which used the Schmidt Telescopes on Palomar Mountain to photograph, monthly, clusters of galaxies in which supernovae might appear. In a single photograph of a cluster of galaxies it was possible to survey many galaxies at the same time. Zwicky and his successor Charles Kowal still hold the records for the number of supernovae discovered by individuals. Similar photographic searches still carry on, on a reduced scale.

Photographic searches provide a permanent record of any supernovae discovered, and the inspection of the photographs can be exceedingly careful, carried out in a comfortable room, perhaps after a sleep. But there is a temptation to postpone the inspection of the photographs after other more urgent tasks are attended to. This may delay the discovery of a supernova, as in the case of SN 1987A, where patrol photographs of the Large Magellanic Cloud taken expressly for finding novae, and showing the supernova at magnitude 6.3, were not looked at; SN 1987A was discovered from another continent a few hours later by eye and on a photograph, the supernova having brightened to magnitude 4.5.

Moreover, the centres of galaxies (where there are lots of stars which have the potential to become supernovae) are bright and saturate the photographic emulsion. Recognition of a supernova in the black mass of photographic grains is very difficult. CCDs (charge-coupled-devices) do not suffer from this problem, but are small and cannot survey large sized clusters of galaxies. There is an alternative: and here amateurs can still play a key role in astronomy.

The eye–brain combination of a human being is an amazingly good, self-adjusting detector which does not readily saturate. It contains a superb image-processing device (a brain). The brain is programmed in a superior but unknown language called *wetware* and it has a memory which, while selective and fallible, enables the brain readily and instantly to recognise changes. Thus it is possible to search for supernovae by eye, looking through a telescope eyepiece.

Robert Evans, an amateur astronomer of Hazlebrook in New South Wales in Australia, is the most successful astronomer to discover supernovae in this way. He has memorised the appearance of over 1000 galaxies near enough to show supernovae if one occurs in them. Each clear dark night he moves his telescope from one to the next, recognising the appearance of a new star in one as if noticing a pimple on the nose of a friend. In six years (10 000 observations per year!) he has discovered over a dozen supernovae.

Astronomers who want to check the appearance of a galaxy to see if the star which they have noticed is a supernova can consult images of galaxies in the *Colour Atlas of Galaxies* by Wray or in charts especially prepared for supernova hunters. It is as well to re-observe the galaxy after half an hour to see if the star is indeed a new one in the field of the galaxy. Minor planets look like stars, and by chance can superimpose themselves on a galaxy, but, while in half an hour a minor planet should move noticeably, a supernova will not!

In an attempt to imitate the human eye–brain combination, astronomers have devised automated searches using CCD detectors and computers. Successful automated searches are run in Berkeley, California, and with the Danish telescope at the European Southern Observatory in Chile. Images of galaxies are obtained automatically and are compared with images obtained previously, one image being subtracted from the other to reveal the new stars. No automated search has yet matched Evans's success.

Once an astronomer discovers a supernova, the other astronomers of the world can

2. *The Australian amateur astronomer Robert Evans holds the record for the visual discovery of supernovae.*

3. *David Malin attempted a colour picture of the supernova discovered by Robert Evans in the galaxy Centaurus A but, being unable at the time of the supernova to obtain a red colour photograph, he used an archive photograph on which no supernova appeared, to complete the trio of red, yellow and blue which he needed. Thus, the supernova, which only appears on the blue and yellow images, shows as an unreal-looking green star in the dust lane which cuts across the galaxy.*

study it, if they are told of its existence. Thus the International Astronomical Union (IAU) has set up an alert network called the Central Bureau for Astronomical Telegrams, which sends telegrams and computer messages about transient phenomena like supernovae to subscribing observatories. The IAU gives designations to supernovae, with the first discovered in a year being given the letter A, the twenty-sixth the letter Z, the twenty-seventh the double letters aa, and so on. To contact the Central Bureau with an alert, telephone Cambridge, Massachusetts, (617) 495 7244, and leave a message on the answerphone. Amateurs in Britain can contact the British Astronomical Association's Deep Sky Section, who will check the supernova and contact the IAU. Amateurs with access to a microcomputer with a 300 baud modem can read about newly discovered supernovae in the latest *IAU Circulars* through the Royal Greenwich Observatory's STARLINK computer by arrangement with the Public Information Unit, Royal Greenwich Observatory, Madingley Road, Cambridge, UK.

Supernova remnants

The explosions which supernovae initiate occur in interstellar space. Whilst more rarefied than the best vacuum achievable on Earth, interstellar space is not completely empty. Typically, in the regions of our Galaxy near our Sun, there is one hydrogen atom per cubic centimetre. The outflow of the supernova's explosion sweeps up this interstellar matter. Indeed, the interstellar material eventually brings the outflow to a standstill, just as the water of an estuary will bring to a standstill the motion of a boat launched down the slipway.

A supernova thus creates a hole in the interstellar material, surrounded by a sphere of

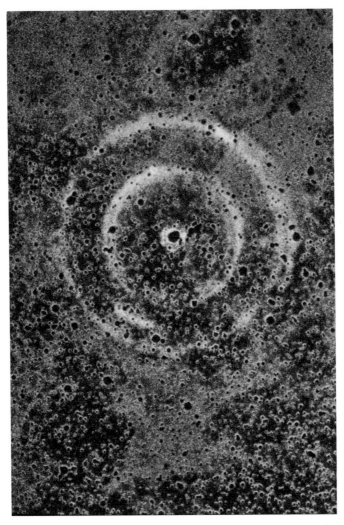

4. *David Malin subtracted the stars and nebulae (black patches and spots) of the Large Magellanic Cloud from pictures taken about a year after the supernova. The image of the supernova is surrounded by concentric rings, echoes of the light flash of the supernova bouncing off hitherto invisible dust clouds.*

compressed matter which is a mixture of the outflowing body of the exploding star and the local interstellar material. These objects, which last for tens of thousands of years, are called supernova remnants, and they can be detected by radio, X-ray and optical telescopes.

Radio telescopes pick up supernova remnants by detecting the synchrotron emission from electrons spiralling in the magnetic field trapped in the shell of compressed star and interstellar material. Radio waves can penetrate right across the Galaxy, uninhibited by dust grains which can conceal optical radiation from supernova remnants from our view from here on Earth. Thus radio surveys for supernova remnants are the most complete.

5. *Pueblo Indians recorded the Crab Supernova on the morning of 4 July 1054 as a bright star near the crescent Moon. This photograph, which shows their record amongst wasp nests in Chaco Canyon, Arizona, is by Miller Goss.*

A recent catalogue of radio supernova remnants lists 135 in our Galaxy. According to their age, they vary between compact complete spherical shells and loose fragments which have all but lost their identity.

In the precise places where radio supernova remnants are detected, it is often (but not invariably) possible to photograph wisps of nebulosity. Sometimes the wisps form complete shells, sometimes only irregular fragments can be seen. If the supernova occurred relatively recently, up to a few thousand years ago, the wisps are the material of the exploding star, the pistons which are driving outwards into the interstellar medium. The supernova remnant is effectively a dissected star, laid open for inspection by astronomers. If the supernova occurred longer ago, the wisps are the interstellar medium, compacted and heated by the shock wave created by the explosion.

Supernova remnants are also known to X-ray astronomers working with satellites to detect X-rays from above the Earth's atmosphere. Most usually the X-ray astronomers detect the heated interstellar material in the shock wave created in the interstellar material by the explosion. Temperatures of millions of degrees are not uncommon.

The first nebula which was identified as a supernova remnant was M1, known as the Crab Nebula. It was first recorded by John Bevis in 1745 and became the first entry in Charles Messier's catalogue of nebulae soon after. It was recognised as a supernova remnant by Knut Lundmark in 1921 and Edwin Hubble in 1928. In 1948 radio astronomer John Bolton tracked down the position of a radio source known as Taurus A well enough to identify it with the Crab Nebula. It proved later to be an X-ray source.

It is one thing to realise that spherical shells seen by different astronomers are one and the same; it is another to prove that the shell was caused by a supernova. Proof of the conjecture came by the identification of some historical supernovae with certain supernova remnants.

In the year 1054 a bright star was seen and recorded by Chinese and Japanese astrologers, and indeed by astrologers of other cultures. The star appeared within a couple

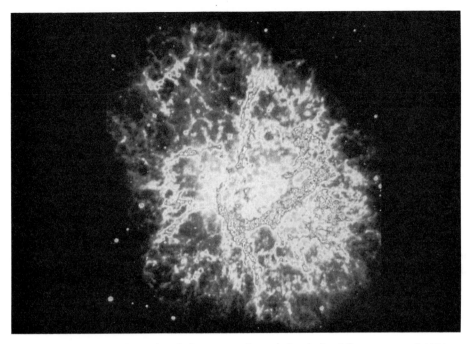

6. *The filaments of the Crab Nebula represent the exploding body of the supernova of 1054, recorded by a scanning CCD on the Kapteyn Telescope on La Palma.*

of days, attained the brightness of Venus, and faded to below naked eye visibility two years later. From the time scale of its light curve it was therefore a supernova. The records of the position of the *guest star* in 1054 (as it was termed by the Chinese) are precise enough to point to the Crab Nebula as its remains. The proof is completed by the measurement of the outflow speed of the Crab Nebula which puts the start of the explosion which created it close to 1054.

The position as well as the brightness of the supernova seen by Tycho Brahe in 1572 were measured carefully by him over its sixteen months' visibility. At the position a complete radio shell and wisps of nebulosity, now known as Tycho's Supernova Remnant, have been found. Kepler's supernova likewise can be identified with a supernova remnant.

Observations of supernovae

The discovery of a supernova triggers astronomers who learn of it to observe it as best they may: occasionally observatories will suspend normal operations, overriding the astronomers who are scheduled at that time, to concentrate on observations of a significant supernova. In this way astronomers have built up a picture of how supernovae behave. Measurements of the brightness of spectra of supernovae over their evolution are typical of the observations made.

Spectra of supernovae show some prominent spectral lines with a characteristic shape known as the P Cygni profile. P Cygni is a variable star in Cygnus; the shape of the

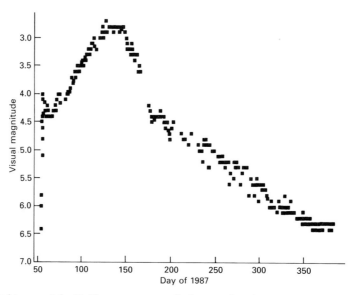

7. *The light curve of the LMC supernova over the first year based upon amateur astronomers' observations as published in the IAU Circulars. The supernova showed a peak magnitude of 2.8, and, in the last two thirds of the year, a linear decline in magnitude at the rate of 0.01 magnitudes per day, due to the decay of the radioactive element cobalt, created in the supernova explosion.*

spectral lines in P Cygni came to be recognised as the signature of an outflowing atmosphere of a star.

The spectra of supernovae not only show that they are outflowing, but they also show the outflow speed. Typical speeds of thousands of kilometres per second are common, with parts of the star outflowing at perhaps up to one-tenth the speed of light. Speeds like this imply the disruption of the star and point to an uninhibited explosion.

If a star is thirty million kilometres in radius and is expanding at 3000 km/s, then it does not take much arithmetic to realise that its radius has doubled in three hours and is one hundred-fold bigger in a week. This astonishing growth in size is the reason why a supernova suddenly brightens. As the star grows in size, however, the light which is radiated from it cools the star, so the supernova eventually fades.

The overall brightening and fading of a supernova form its light curve; some light curves of the brighter supernovae have been well recorded by amateurs from eye observations. The light curve of SN 1987A is a good recent example. Southern hemisphere amateurs followed the supernova, comparing it with standard stars of known brightness, at first with their unaided eyes, then with binoculars and telescopes.

In producing this light curve, amateur astronomers were imitating the observations made, collected and analysed by Brahe in 1572 about his supernova and by Kepler in 1604 about his. These two supernovae had light curves which were very similar and distinctively different from the light curve of the Large Magellanic Cloud supernova. Astronomers recognise that the light curves of about one-half the supernovae discovered

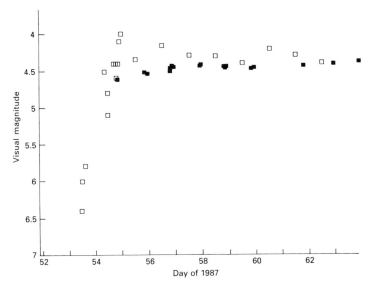

8. *The initial days of the light curve of the LMC supernova were recorded by amateurs' measurements (light squares), confirmed in later days by photoelectric measurements by professionals (filled squares). The amateur measurements show a small overshoot in the second day of the light curve compared with the professional measurements – the eyes used to observe the supernova responded to its blue light which it was emitting copiously in its first day or two, when it was hot.*

are of a similar type, unimaginatively called Type I. The rest are Type II. Supernovae which have Type I light curves have spectra which are also very similar to each other; in particular the spectra have no indications that hydrogen is present in the explosion. Supernovae whose spectra do show hydrogen are called Type II.

The absence of hydrogen in the spectra of Type I supernovae is remarkable, since hydrogen is by far the most common element in the universe and practically everything contains it.

The parent galaxies of supernovae

The fact that supernovae occur in galaxies outside our own makes it possible to determine the energy released in a supernova explosion. The brightness of a supernova at its maximum can be related to the distance of the parent galaxy through its red-shift.

It turns out that Type I supernovae have brightnesses at maximum which are the same from one Type I supernova to another; the jargon phrase which astronomers use to describe this is that Type I supernovae are *standard candles* – luminous objects whose powers are all the same. Astronomers can in principle use supernovae in distant galaxies to measure their distances and check the relation between distance and red-shift at cosmological distances, to determine properties of the early universe (although this project has never successfully been completed).

Type II supernovae are more of a rag-bag of miscellaneous objects – 'non-standard candles', if you will. But they are on average somewhat less powerful than Type I supernovae.

Rather surprisingly, Type I and Type II supernovae occur in different kinds of galaxies. Type II supernovae occur only in galaxies with nebulae, star clusters and, usually, spiral arms. They never occur in elliptical galaxies, in which these phenomena are absent. Type I supernovae, on the other hand, occur everywhere, in spiral galaxies and elliptical galaxies alike.

Type II supernovae

These subtle but distinctive properties suggest that there are two kinds of stars which explode as supernovae.

Astronomers believe that Type II supernovae occur at the ends of the lives of ordinary but massive stars – these are the kinds of stars which give spiral galaxies their recognisable appearance, with star clusters (which include massive stars) and nebulae (which are illuminated by them) forming the spiral arms. There are no massive stars in elliptical galaxies since few stars have recently formed in elliptical galaxies, and the massive stars formed in them a long time ago have all rapidly evolved and died long since. This is why there are now no Type II supernovae in elliptical galaxies.

Until 1987, the link between Type II supernovae and massive stars was only a deduction from subtle evidence. In 1987 SN 1987A was observed to explode precisely at the place

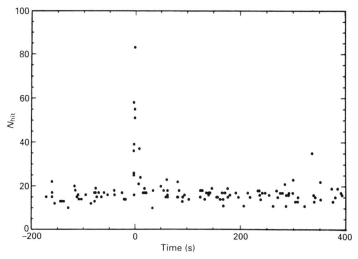

9. The pulse of neutrinos from SN 1987A was caught in the Kamiokande detector as a unique spike in its records. A ten-minute segment of its record about 07:35 GMT on 23 February, 1987 (t = 0 in this graph) shows the number (N_{hit}) of photomultipliers surrounding the tank of water which were active from moment to moment. Usually between ten and twenty photomultipliers were showing signals from photons generated in the water by cosmic rays and radioactivity but at certain moments near t = 0 a large number (up to ninety) of the photomultipliers were hit by photons from Cerenkov radiation created in the tank by neutrinos from the supernova. About seven neutrino events were recorded.

Paul Murdin

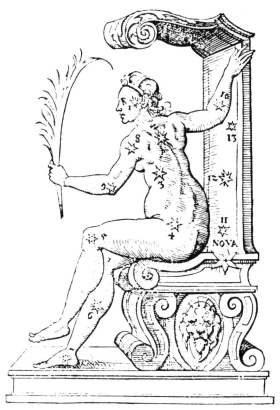

10. *Tycho's supernova of 1572 completely altered the W-shape of Cassiopeia, as shown by Czech astronomer Thaddeus Hagecius who, himself, observed it.*

where a massive star had been; the star had been catalogued and named as Sk-69 202. As SN 1987A faded, it became clear that Sk-69 202 had disappeared. For the first time, a massive star had been observed to explode as a Type II supernova.

The crucial stage at the end-point of the life of a massive star which triggers a Type II supernova explosion is that its centre runs out of nuclear fuel. Stars shine by creating energy from nuclear fusion, with hydrogen nuclei fusing to form helium nuclei, helium nuclei fusing to carbon nuclei, carbon to oxygen, etc. The fusion chain ceases when iron nuclei are formed and there is no more fusion energy available. The flow of energy from the core of the star is what keeps the star up against the force of gravity. When the nuclear fusion chain ceases, no more energy flows up in the star, there is no pressure force to balance the star's own weight, gravity wins, and the core collapses to form a tiny neutron star, or possibly a black hole.

Energy is released in the fall of the core, heating it and its immediate surroundings to enormous temperatures (30 000 million degrees). The energy is radiated from the core, principally as neutrinos, elusive nuclear particles, and as heat. Some (1% or so) of the radiated energy is caught by the outer parts of the massive star, overcomes the force of gravity binding the outer parts to the core, and ejects the outer parts in a gigantic

149

explosion. What we see as a supernova explosion – heat, light, kinetic energy – is, although enormous, just a percentage of the total energy which is released. The exact amount which we see radiated from a supernova of Type II depends on the nature and size of the star whose core collapsed; this is why Type II supernovae are such a rag-bag of miscellaneous objects.

According to theory, the collapse of a stellar core produces a neutron star. This is a star so small (radius say of 15 km) that the material of which it is made is in the form of neutrons rather than electrons, protons and neutrons as in the case of normal sized stars like the Sun. The density of neutron stars is the same as the density of the atomic nucleus – densities of 100 million tonnes per cubic centimetre are typical.

Neutron stars rotate quickly, especially when they are first formed, and they have strong magnetic fields. They manifest themselves as pulsars, the rotation of the neutron star causing a lighthouse-like beam to sweep repetitively through space and be seen as the regular pulses of light, radio or X-rays, which give pulsars their name. There is a pulsar and therefore a neutron star at the centre of the Crab Nebula, the remnant of a Type II supernova. (Judging by the hydrogen in the spectrum of the remnant, the supernova which caused it contained a lot of hydrogen and the supernova of 1054 would therefore have had a Type II spectrum.) There is another neutron star at the centre of the Vela supernova remnant, which judging by its location amongst a lot of massive stars was also a Type II. It seems that Type II supernovae do indeed, on at least some occasions, generate neutron stars.

SN 1987A in the Large Magellanic Cloud was a Type II supernova (its spectrum had hydrogen lines). There is direct evidence that a neutron star formed in the supernova explosion, since, on 23 February, 1987, scientists detected from the supernova a burst of neutrinos, elusive subatomic particles. The properties of the neutrino burst were in good agreement with calculations of the burst expected from the formation of a neutron star. Two reports that the flashes of a pulsar had been seen in the debris of the explosion have never been confirmed and there is no direct or indirect evidence of its existence. This is a puzzle.

Some supernova remnants look like Type II supernovae in that they are made of wisps of material which seem to have the composition of an exploded massive star; but there is no evidence either direct or indirect that they have pulsars. Cassiopeia A is such a supernova remnant, and it is conjectured that it contains a black hole, but no evidence is expected from this.

Black holes, and pulsars which have lost their energy to become dead neutron stars, contribute to the dark matter which populates the space between the visible stars. It is very hard to estimate the amount which they contribute but, it seems, they do not add up to enough to solve the problem of the *missing mass*.

Type I supernovae

Most astronomers believe that Type I supernovae, in contrast to those of Type II, occur as the explosions of white dwarf stars, which are commonly found everywhere in all sorts of galaxies, having been created at the ends of the lives of the less massive stars.

White dwarf stars can be made of various materials, depending on how they were formed. The most popular theory of Type I supernovae is that they are explosions of carbon–oxygen white dwarfs; this is why Type I supernovae have no hydrogen in their spectra. There are two (at least) scenarios which astronomers have invented to explain Type I supernovae.

One theory is that, for one reason or another, a carbon-oxygen white dwarf forms in a binary system, circled by another star. This star evolves, grows larger, and leaks its atmosphere onto the white dwarf, whose weight thus increases. It increases to a point beyond which the white dwarf cannot keep itself up. It collapses, releases lots of energy (neutrinos and heat), and blows itself apart.

Another theory is that, again for one reason or another, two white dwarfs form in a binary system, each individually less than the critical value for their weight. They lose energy and grow closer together, eventually merging and exploding as before. This scenario is especially interesting since the final pursuit and plunge of the two stars together creates gravitational radiation, which might, with foreseeable advances in technology, soon be within the range of a terrestrial gravitational wave detector.

Supernovae of Type I, according to these theories, create no neutron stars nor therefore pulsars; there are none at the centres of the remnants of Brahe's or Kepler's supernovae, which, to judge by their light curves, were Type I, so theory is in agreement with what we know.

Creation of the elements

The densities and temperatures in supernovae are high enough to cause nuclear reactions to occur as the matter of the star which is exploding is cooked. Large amounts of the common nuclei are created this way (as well as in other stellar processes). These nuclei form the common elements of your surroundings at this moment, the carbon and oxygen of the book which you hold, the silicon and oxygen of the building material enclosing you. In addition supernovae uniquely create some rare nuclides: the gold of your ring or watch, for example.

Some of the nuclides created in the supernova explosion are radioactive. In SN 1987A about 0.07 of a solar mass of the nuclide nickel-56 was created. It suffers beta-decay, and changes first to cobalt-56 and then to iron-56, which is stable. In the spectrum of the supernova, astronomers watched the spectral lines of cobalt weaken and those of iron strengthen as this change happened.

In the radioactive decay of nickel and cobalt, energy is given off in the form of gamma rays and energetic electrons. The gamma rays are emitted at particular energies. Such gamma rays from SN 1987A were detected by the satellite Solar Max, the first time that gamma rays had been definitely identified from a celestial source outside the solar system.

The energy from radioactivity powers the supernova after the first initial release of energy in the collapse of the progenitor star; before the envelope of the star gets so large and tenuous as to become transparent, the radioactive decay energy is trapped in the expanding envelope of the supernova and is re-processed to light and infra-red radiation. The amount of energy which is radiated in this way, and results in the supernova's light

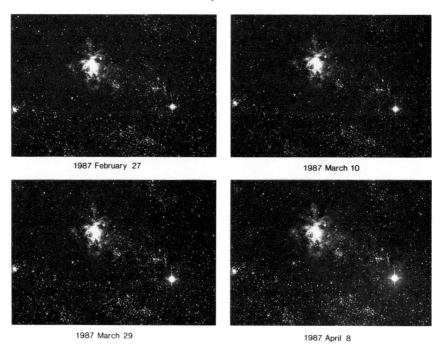

1987 February 27

1987 March 10

1987 March 29

1987 April 8

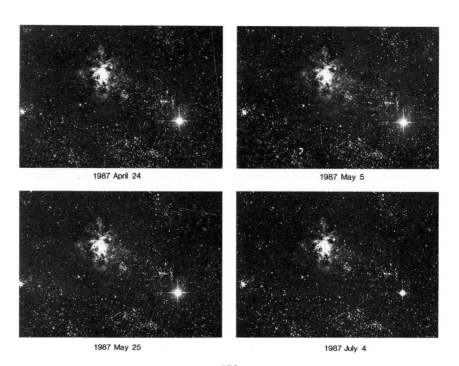

1987 April 24

1987 May 5

1987 May 25

1987 July 4

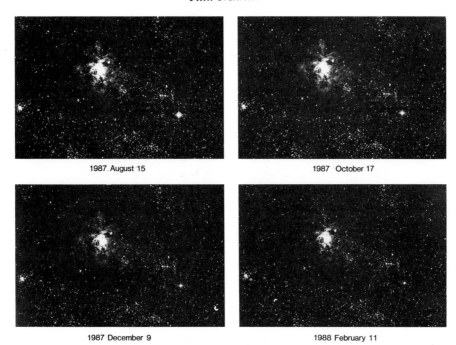

1987 August 15

1987 October 17

1987 December 9

1988 February 11

*11. The change in magnitude of the Large Magellanic Cloud supernova during the first
year is recorded in a collection of photographs from the UK Schmidt Telescope in Australia.*

curve, can be directly related back to the radioactive decay of cobalt-56; those amateur
astronomers who watched SN 1987A fade, magnitude by magnitude, after about August,
1987, when the initial effects of the supernova explosion had faded, were seeing radioactivity on a stellar scale.

These nuclides, as well as those unprocessed in the supernova explosion but forming
the body of the exploding star, are dispersed by the force of the supernova explosion, at
first to form part of a supernova remnant, and then, after millions of years, to dissipate
into space, mixing with the interstellar material and, perhaps, eventually, forming new
stars and even planets. Indeed, long past, never-to-be-identified supernovae created nuclei
which mixed with the solar nebula of 5000 million years ago and formed our solar system,
our Earth, our bodies. Man is himself a supernova remnant.

The history of our Galaxy is, in part, controlled by supernovae. The explosions of
supernovae produce an increasing supply of the elements made in stars, which mixes with
the material made in the big bang at the start of the universe (principally hydrogen and
helium). Older stars, which formed early in the history of the Galaxy, contain less of the
elements other than hydrogen and helium than stars which have formed more recently,
from heavily 'polluted' material. There is evidence in the way that the composition of the
Galaxy changed that there was a wave of supernovae early in its history, when supernovae
were occurring every year rather than every century.

Supernovae not only produce changes in the composition of our Galaxy, their
explosions also transport material from place to place in the Galaxy. Type II supernovae

are from massive stars which populate the spiral arms, lying in the gas of the galactic plane. Supernova explosions compress this gas and trigger the formation of further generations of stars. They also send 'fountains' of gas into the halo of the Galaxy, circulating gas into regions in which it is not usually found.

Supernovae are rare, but their powerful explosions have shaped the destiny of the solar system, mankind and the Galaxy itself.

11

The story of the pulsars

FRANCIS GRAHAM-SMITH

Neutron stars

For most of us the stars are the bright points of light we see in the night sky, or the fainter points of light we study with optical telescopes. There are, however, some stars which are invisible, and which we can study only because they emit radio waves or X-rays. Among these are the pulsars. They are very small stars, only one-hundred-thousandth of the diameter of a normal star like the Sun, and yet they contain about the same mass as the Sun. Their densities are greater than that of the Sun by a factor of a thousand million million. How can such compressed and dense objects exist, and how can they be formed?

On Earth the same high densities occur in the central nuclei of ordinary atoms, where protons and neutrons are tightly packed together. The pulsars are like giant atomic nuclei, but containing only tight packed neutrons. Such objects are the neutron stars, only 20 km across, so small as to be virtually invisible, and at the same time such powerful sources of radio waves that over 400 have been discovered by radio astronomers.

The idea of a neutron star goes back to 1932, the year of the discovery of the neutron itself. When the news of James Chadwick's discovery of the neutron reached Neils Bohr's Institute in Copenhagen, the Russian physicist Lev Davidovich Landau was there. He immediately pointed out that a neutron star was a theoretical possibility, but it was not thought that such an object could ever be observed. Even if it were very hot, say at a

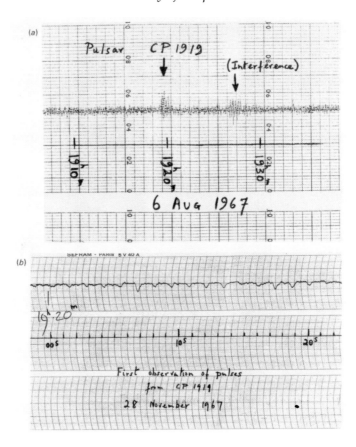

1. *Discovery observations of the first pulsar.* (a) *The first recording of PSR 1919 + 21; the signal resembled the radio interference also seen on this chart.* (b) *Fast chart recording showing individual pulses as downward deflections of the trace.*

million degrees kelvin, the radiation from such a tiny object would be undetectable at stellar distances.

The possibility that such a star might be formed in a supernova explosion was evident to Walter Baade and Rudolf Minkowski, who followed Landau's remark with a paper suggesting that such an explosion might be the result of the collapse of a star at the end of its normal life. The collapse would release a tremendous burst of energy, seen as an expanding cloud of hot gas, but part of the mass would remain as a neutron star. The explosion could also provide the answer to two of the most important questions in astrophysics: what was the origin of the heavier elements, and what was the origin of the cosmic rays? Baade and Minkowski were very close to the discovery of the neutron star that we now know as the Crab pulsar, which they knew as a very strange star but which could not, according to the theory of neutron stars, possibly emit so much light.

Francis Graham-Smith

The discovery

In the event it was not the light emitted by a neutron star, but its radio emission, that allowed its discovery by Antony Hewish and his student Jocelyn Bell in 1967. They were using a specially constructed radio telescope to monitor the fluctuations in the radio signals from distant radio sources, and especially the fluctuations resulting from ionised clouds in the line of sight, such as the energetic clouds thrown out from the Sun. They found a fluctuating radio signal which was nowhere near the Sun, and then discovered that its signal was a very regular series of pulses, at intervals of 1.34 s. No theory of neutron stars had predicted such a phenomenon, and many different explanations were explored, such as an oscillating white dwarf star, or even some form of extraterrestrial intelligence. They had, as it turned out, observed a rapidly rotating neutron star. Within weeks several others were found, and the new class of objects were named 'the pulsars'.

A much faster pulsar, discovered by Australian radio astronomers, gave two clues that the source of the pulses was not an oscillation but a rotating 'lighthouse' beam. This was the Vela pulsar, with a period of 89 ms. Its pulses were found to be highly plane polarised, with a steady sweep of position angle which was easily interpreted as radiation from diverging magnetic field lines over a magnetic pole. Second, the period was steadily increasing, as expected from a rotating star which was losing energy and angular momentum.

The Crab Nebula and its pulsar

The Crab pulsar, discovered at the Arecibo radio telescope, provided the crucial link with Baade and Minkowski's perceptive suggestions. This pulsar has a period of only 33 ms, and its very rapid rate of slowdown shows it is only about 900 years old. Its age, and its location in the Crab Nebula, shows that it was born in the supernova of AD 1054, which was observed and recorded by Chinese astronomers. Furthermore, the rate of slowdown meant that it was losing enough energy to provide for the whole of the emission from the surrounding Crab Nebula. Until then there was no explanation for the continuing radiation from the Nebula, especially the high-energy X-radiation which would fade completely within a few years without a continuous supply of energetic charged particles. This energy comes from the strong magnetic field of the neutron star, of order 10^{12} gauss on the surface, which radiates a 30 Hz electromagnetic wave as the star rotates.

The long-standing mystery of the visible star in the centre of the Crab Nebula was solved shortly afterwards when the light from this star was found to be pulsing at 30 Hz. The radio pulses and the light came from the same object, the neutron star remnant of the supernova explosion. This was the first time that a pulsar was identified with a visible object. The discovery that the high-energy radiation extended beyond the optical range into the X-ray and gamma-ray spectrum followed soon afterwards, showing that the pulsar spectrum extended over a range of more than fifty octaves.

157

2. *The Vela supernova of about 10 000 years ago left this wonderful fine nebula in the southern sky.*

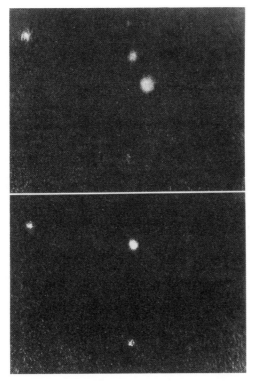

3. *The Crab pulsar. This pair of photographs was taken by a stroboscopic technique, showing the pulsar on (above) and off (below).*

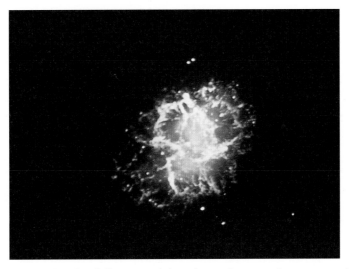

4. *The Crab Nebula. This shell of gas and the pulsar within it are the remains of the supernova of 1054.*

The galactic population

Only four pulsars which emit both radio and X-ray pulses are known. Most emit radio only, and over 400 of these are now known. Radio surveys designed to detect pulsars have covered the whole sky, and the presently known sample is sufficient for a reasonable statistical analysis of their distribution in the Galaxy. The total number in the Galaxy is believed to be about 100 000; most are too distant for their signals to be detectable, and more than half have radio beams that point away from our line of sight. A much larger sample is very hard to obtain. The surveys can only be carried out using the largest radio telescopes; even then a sophisticated analysis technique is needed to pick out a very weak, but very precisely periodic, signal (of unknown period) from the predominant receiver and sky noise. Some parts of the sky have been searched with special attention; in particular the Magellanic Clouds have each yielded one pulsar, the most distant so far recorded.

The observable rate of evolution, measured by the slowdown of rotation, shows that most pulsars are only observable for about ten million years, after which time they have slowed down to a periodicity of several seconds and have stopped radiating. We believe, although we are not sure, that they all originate in supernovae, which generally occur near the plane of the Galaxy. They cannot travel far in their lifetimes, so they are mostly found close to the plane. The actual distances travelled depend on their velocities, which may be up to 300 km/s. These high velocities are probably related to an origin in disrupted binary star systems, as will become clear when we turn to the very short-period pulsars, the millisecond pulsars.

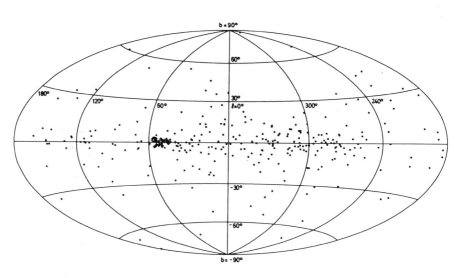

5. *Distribution on the sky of 316 pulsars. These constitute a sample over the whole sky down to a well-defined flux limit.*

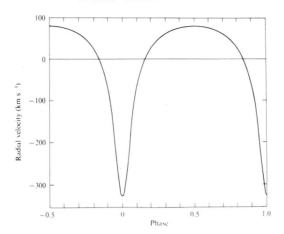

6. *Radial velocity curve for the binary pulsar PSR 1913 + 16. The velocity is found from the modulation of the pulse period due to the Doppler effect. The curve is markedly non-sinusoidal, indicating the large eccentricity of the orbit.*

Millisecond and binary pulsars

The simple picture of a pulsar life history: born in a supernova, slowdown for ten million years, and then fading to insignificance, was complicated by the discovery in 1982 of the first of the millisecond pulsars. This pulsar, PSR 1937 + 21, was already known as a radio source with the steep spectrum and high polarisation characteristic of pulsars, but no pulsations could at first be found. After several searches for pulsations within the normal range of some tens of milliseconds upwards, it was found to be a pulsar with a period of only 1.5 ms. This might have indicated that it was very young, but it was not, as would be expected, rapidly slowing down. On the contrary, it was evolving only very slowly, with a lifetime at least 1000 times greater than normal. More than ten such millisecond pulsars are now known, most having been found in similar targetted searches of individual objects or locations. The most fruitful regions for these searches have been the globular clusters. The majority of the millisecond pulsars are in binary systems, with another condensed star, either a neutron star or a white dwarf, as a companion.

The majority of the stars in the Galaxy are not solitary, but in orbit around companion stars, forming binary or multiple systems. In contrast, most pulsars are solitary. It is immediately obvious when they are in orbit around a companion, whether visible or not, from the varying Doppler shift of their pulse period. An especially interesting binary is PSR 1913 + 16, in which the binary period is only $7\frac{3}{4}$ hours. The unseen companion here is another neutron star. The orbital motion is so rapid that its analysis involves full-scale general relativity: in fact, it provides the modern text-book example of several effects including orbital precession and gravitational radiation.

The binary systems are the key to the origin of the millisecond pulsars. At the end of the life of a normal pulsar, its rotation rate and its magnetic field have both decayed. If, however, it is in a binary orbit, its partner may be able to rejuvenate it by spinning it up to a much faster rate than it previously enjoyed as a young pulsar. It achieves this by

7. *The globular cluster 47 Tucanae which contains a remarkable pulsar. The pulsar is in a binary system with an orbital period of only 32 minutes.*

transferring some of its mass, which falls on the old pulsar, carrying some of the large angular momentum of the binary orbit. This occurs when the companion evolves into a giant star, whose envelope spills over into the gravitational pull of the neutron star. The speed up can take the old pulsar to such a speed that it is nearly bursting under the centrifugal force; the fastest so far known is rotating at 647 times per second. After the transfer, the rotation rate only decreases very slowly, since the magnetic field has decayed by a factor of 1000. The same process of mass transfer also accounts for the pulsating X-ray sources in the Galaxy; the energy released when the material falls on the neutron star heats the infalling gas to temperatures of ten million degrees or more, making a binary system which is among the strongest X-ray sources in the sky.

Interaction between stars is obviously more common where the concentration of stars is greatest. This is certainly true in globular clusters, where binary X-ray sources and their products, the millisecond pulsars, are often to be found. A most remarkable pulsar in the globular cluster 47 Tucanae is in a binary system with an orbital period of only 32 minutes, the shortest period for any orbiting system so far known.

Since most stars are in binary systems, it is surprising to find that most of the pulsars are solitary. How did these solitary pulsars lose their partners? Several theories have been proposed. One possibility is that the partner itself exploded as a supernova, losing so much material that the binary was disrupted. Another is that the partner was a less massive star, possibly a white dwarf, in a close orbit round the energetic pulsar. In this case the partner might be so bombarded with radiation that it was evaporated. There is a binary pulsar system in which this extraordinary process can be seen; the partner has a mass of only 0.05 solar masses, and it is surrounded by a cloud of hot material, apparently evaporated by gamma rays from the pulsar.

Glitches and timing noise

Millisecond pulsars are generally very good clocks. After allowance is made for the regular and predictable slowdown of their rotation, the radio pulses from a solitary millisecond pulsar arrive in a time sequence which is accurate and predictable over long periods to about 1 μs. If several such pulsars were regularly monitored, they could provide a better time scale than any terrestrial clock. Other pulsars are less predictable, showing random wandering and discrete steps in the pulse arrival times. Eight such steps, known as glitches, have been observed in the Vela pulsar during twenty years; in each the steady slowdown of the pulsar was interrupted by a sudden speed up of about one part in a million. Glitches are seen in pulsars and the whole range of periods, from the Crab pulsar with 33 ms period through to the longest period of 4 s. The Crab pulsar itself only has small glitches, in which the period changes by one part in 10^9; however, it shows in addition a wandering of pulse arrival time which is quasi-periodic with a period of about eighteen months.

These irregularities in rotation originate deep within the neutron star. The outer parts

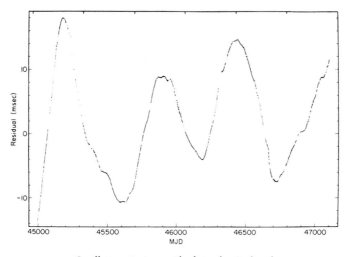

8. *Oscillatory timing residuals in the Crab pulsar.*

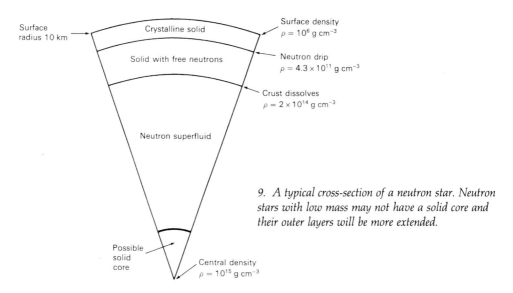

Surface radius 10 km

Crystalline solid

Solid with free neutrons

Neutron superfluid

Possible solid core

Surface density
$\rho = 10^6$ g cm^{-3}

Neutron drip
$\rho = 4.3 \times 10^{11}$ g cm^{-3}

Crust dissolves
$\rho = 2 \times 10^{14}$ g cm^{-3}

Central density
$\rho = 10^{15}$ g cm^{-3}

9. *A typical cross-section of a neutron star. Neutron stars with low mass may not have a solid core and their outer layers will be more extended.*

of the star form a solid crust, an extremely strong crystalline structure of iron nuclei. The interior, consisting of neutrons with a small proportion of electrons and protons, is a liquid which is both superfluid and superconducting. The fluid rotates at nearly, but not quite, the same speed as the crust. The coupling between the fluid interior and the crust changes, either discontinuously in a glitch or more smoothly in the timing wander seen in the Crab pulsar. There is here a close link with laboratory physics, with similarities between the behaviour of the neutron fluid and liquid helium, which is also a superfluid. The rotation of a superfluid is different from that of normal fluids: instead of the fluid rotating as a whole, it contains discrete vortices, each representing a unit of rotation. As the rotation of a pulsar decreases, so the density of vortices must decrease, and the individual vortices must move outwards. The glitches, and the oscillation, occur as a result of a tendency for the vortices to become attached to the individual nuclei of the crystal lattice. A complete pinning of all vortices would prevent the outward flow, so that the interior could not slow down. The glitch is then a release of vortices which have been caught up in this way, while the oscillation is a differential movement between the centre and the outside of the neutron fluid.

The pulsar magnetosphere

Without a magnetic field, any atmosphere outside a neutron star would be condensed by the enormous gravitational field to a layer 1 mm thick. The magnetic field is, however, so strong that the rotation generates an electric potential difference between pole and equator of up to 10^{15} V. This pulls charged particles out of the surface, forming a dense and energetic magnetosphere. Without this there would be no appreciable radiation, either radio or visible. The magnetosphere is much larger than the pulsar itself. It rotates with the neutron star, and it extends out to such a distance that the outer edge is moving with almost the speed of light. The dipole magnetic field of the star dominates the physical conditions in the magnetosphere, and determines the locations of the radio and optical emission: both are on magnetic field lines originating near the magnetic poles, the optical source being near the outside edge of the magnetosphere, while the radio is located closer to the surface. The radiation mechanism of the optical radiation is well known in the laboratories of high-energy physics; it is the same as the synchrotron radiation from the charged particles in high-energy accelerators. The radio is less well understood; it is an instability in a plasma whose high density and relativistic energy are outside any of our laboratory or theoretical experience.

Probes of interstellar space

Pulsars are sufficiently fascinating in themselves for a lifetime of research. They can, however, be used as tools for several other astrophysical investigations concerned with the interstellar medium. The fluctuations seen by Hewish in the signals from the first pulsar were in fact not intrinsic to the pulsar, but a scintillation effect due to the electron clouds on the radio propagation path. The rate of fluctuation depends on the velocity of the pulsar across the line of sight, and on the radio frequency: the lower the frequency, the faster the scintillation. There is also a slower scintillation, referred to as refractive rather than diffractive scintillation, which is due to larger scales of electron clouds; this may have a time scale of months rather than minutes.

Pulsars give a measure not only of the variations of electron density; they also give a measure of the total content of electrons along the line of sight. This is found from the delay in pulse arrival time at lower radio frequencies. The pulses may be delayed by several seconds in their journey time of some hundreds or thousands of years; the delay depends on radio frequency, and is quantified as a dispersion measure (DM). It turns out that the average electron density does not vary much over a large part of the Galaxy, so that the DM can be used as a good measure of pulsar distances.

Finally, pulsars can be used to measure the magnetic field in interstellar space. Although this field is very weak, typically only a few microgauss (1μG is 10^{-10} tesla), it acts with the electrons to rotate the plane of polarisation of radio waves. This Faraday rotation can then be combined with the measured DM to determine an average value of the line-of-sight magnetic field. Since pulsars can be found in all directions and at distances up to and beyond the galactic centre, these measurements are uniquely useful. They show that the magnetic field follows the direction of the spiral arms, as expected; the surprise is that the direction reverses in the next spiral arm, closer to the centre of the Galaxy.

12

The Milky Way Galaxy

HEATHER COUPER AND NIGEL HENBEST

To the ancients, the hazy band of light that spans the sky on a clear night held no mysteries. If anything, their explanation for it was somewhat down-to-earth. They told the legend of Hercules, one of Jupiter's illegitimate offspring, who was desirous of immortal life. Unfortunately for him, his mother was a mere mortal. Jupiter took pity on the child, and informed him that suckling at the breast of a goddess would ensure immortality. So Hercules crept up to the sleeping Juno – but the goddess was woken by his approach, and the stream of milk splashed across the sky. So was created what was literally the Milky Way.

Despite the ancients' delightful certainty, only this century was the Milky Way's nature fully realised. Ever since Galileo first turned a telescope to the sky in 1609, astronomers had known that the Milky Way was made of stars, but they were unsure how to reconcile these distant and apparently crowded stars with the brighter stars that we can pick out with the naked eye. Eighteenth-century philosophers like Thomas Wright and Immanuel Kant hazarded commendably accurate guesses, but most professional astronomers of that time were busy studying the solar system.

It was left to an amateur astronomer, the great William Herschel, to tackle the question scientifically. Using his large home-made telescopes, he carefully counted the number of stars visible in different directions, to see just how they were distributed in space. He

1. *The Origin of the Milky Way by Tintoretto. Milk from Juno's breast spurts upwards to form the Milky Way. More milk rains down on Earth to form the lilies. The rest is drunk by the infant Hercules who is being supported by Jupiter.*

even hit on the correct explanation of the layout of the Milky Way, but later abandoned this idea.

By the start of the twentieth century, however, enough strands of evidence had come together to show that the band of the Milky Way is just an edge-on view of the distant stars that make up our home Galaxy. The discovery of millions of other galaxies spread throughout space — each a separate star-island with a membership running into millions — also helped astronomers to gain a perspective on our own.

With hindsight, the broad-brush picture of our Milky Way Galaxy is surprisingly simple. It is a disc-shaped system of stars, gas and dust wheeling slowly in space once in 200 million years. From the side, it would look like a spindle, widening towards the central bulge. A bird's-eye view would reveal the beautiful Catherine-wheel shape of a typical spiral galaxy, so wide that a ray of light would take 100 000 years to cross it.

From intergalactic space it would be very difficult to spot the Sun, for it is just one out of the Galaxy's 200,000 million stars, and not a very brilliant one at that. The Sun is situated in an unceremonious position in the galactic suburbs some two-thirds of the way out from the centre. We live in the outer part of the Galaxy's disc, and it is this positioning that gives rise to the band of the Milky Way in our skies. Looking into the thickness of the disc all around us, there are stars as far as we can see, tailing off into a misty blur. These distant stars appear to be crowded together, but this is only an effect of perspective. When we look above or below the disc, the stars quickly run out: beyond is just empty space.

The disc is the most obvious part of the Galaxy, but it is far from being the only component – or even the most important part. It is surrounded by a giant 'halo', a vast spherical region some 300 000 light-years across, which is the ghost of the original gas cloud that collapsed to form the Milky Way. Now the halo is populated by stars so old and dim that it is difficult to see them at all – except where they congregate into the aptly named 'globular clusters': spherical collections of stars containing up to a million stars each. These are the Galaxy's first citizens; the stars that formed some 13 000 million years ago, when a cloud of gas fresh from the big bang began to turn into a galaxy.

But the ghostly halo may be more substantial than it seems. The motion of the gas and stars in the disc of our Galaxy suggests that they are under the influence of gravitational forces far stronger than could be mustered by the visible matter in the

2. From Earth our Galaxy looks like a band of stars around the sky. This river of light is known as the Milky Way. This view was photographed by Akira Fujii.

3. *Our Galaxy is a disc-shaped system of stars, gas and dust. Edge-on its Catherine-wheel spiral shape is evident. The Sun is situated off centre, in one of the spiral arms. From the side the Galaxy is shaped like a spindle, widening toward the central bulge.*

Galaxy. There must be a huge amount of 'dark matter'. We cannot detect the dark matter with any kind of telescope that exists at present, and it makes its presence felt only by its gravitational pull. According to recent calculations, we see only one-tenth of the Galaxy's total mass; the remaining nine-tenths is the dark matter that resides, presumably, in the halo.

What could make up this dark matter? It would be convenient if it took the form of gas left over from the Galaxy's formation; but the halo is, in fact, more or less free from gas of any kind. At the moment, astronomers are investigating several very different possibilities. The halo may be chock-a-block with some kind of elementary particle: these could be something we know about already, like neutrinos (which would then have to possess a certain amount of mass, contrary to some experiments in the laboratory), or particles that physicists have predicted theoretically but have yet to find in nature. These particles rejoice under names like axions and gravitinos. Or there may be huge numbers of individual astronomical objects that produce very little light or other radiation. Black holes spring immediately to mind, as they are the ultimate in very dark objects. But a multitude of objects like Jupiter would do just as well. Planets floating freely in space would produce practically no radiation of their own.

The disc of our Galaxy is much better understood than the halo. It is the Galaxy's newest neighbourhood. Its star population contains a fair sprinkling of older inhabitants, but many of these are relegated to its outer fringes, above and below the galactic plane. In fact, the whole distribution resembles a series of tide marks, with stars of younger and younger age marking the stages of its formation, as it collapsed from a fat gas cloud to a thin disc. The central plane of the Galaxy is marked by the youngest stars: it is the region where the stars of the Galaxy are still forming.

Here the spaces between the stars are filled with invisible, tenuous gas, the raw material of future stars. There are only half a dozen atoms in a matchboxful of space, on average, but in time these isolated atoms – mainly hydrogen and helium, but increasingly atoms of carbon, nitrogen and oxygen processed in stellar furnaces and ejected into space – will clump together in huge clouds to produce the next generations of stars. The immediate products of starbirth are all around us: we are lucky enough to live in a region of the Galaxy where our skies are bright with young nebulae, sparkling star clusters and dazzling, but short-lived, blue supergiant stars.

These spectacular youngsters have a scientific use too: they allow us to investigate the geography of the Galaxy. We know, by investigating other spiral galaxies, that these kinds of objects are closely grouped into spiral arms, while older stars are spread more evenly over the disc. When we measure the distances to the young objects near the Sun, and plot them out on a map of the Galaxy, we find that they bunch into three distinct bands – one at about the Sun's distance from the galactic centre, one closer in, and the other rather further out. These, without a doubt, mark the positions of the three nearest spiral arms of our Galaxy. The Sun is a member of the Orion Arm – along with the bright nebulae and young stars that make up the striking figure of Orion, the hunter. It is flanked by the Sagittarius Arm, closer to the galactic centre, and the Perseus Arm farther out.

Optical telescopes give us a grandstand view of the inhabitants of these arms. But it is difficult to work out what is going on further away in our Galaxy. The problem is not just that objects look fainter when they lie at a greater distance, but that our vision is also clouded by particles of dust in space. These confine our view to a region only 10 000 light-years across – only one-tenth the size of the Galaxy.

At first, dense regions of dust seemed to be a veritable blessing. These dark clouds (like the Coal Sack near the Southern Cross) look to the naked eye like starless voids, and many nineteenth-century astronomers thought them to be 'tunnels' through which we could peer vast distances into space. Ironically, these dark regions are in fact dense patches of dust which block off the light from more distant objects.

These dark clouds are the densest regions of 'fog' in the Galaxy; but the dust particles are spread out more thinly everywhere else in the disc, to produce a general, if only slight,

4. *The dark clouds of the Coal Sack (centre) are silhouetted against the background of the Milky Way. Just above and to the right are the four brilliant stars that the form the cross of the constellation Crux. The two bright stars at left are Rigil Kentaurus and Hadar, in the constellation of Centaurus.*

interstellar 'mist'. On average, this dims a star by one magnitude (reducing its light by 60%) for every 3000 light-years of its distance. But the dust is far from uniformly spread, and the amount of obscuration is unpredictably uneven.

Most astronomers believe that this 'dust' consists of small particles of rock (silicates) and soot (carbon), coated with ice, and each less than a micron across. They condensed from old red giant stars, which are continually losing gases from their surfaces into space. Although it can be prominent in the dense clouds, the dust makes up only 0.1% of the Galaxy's mass.

As far as the fog-bound astronomers on Earth are concerned, the grains of dust have one very big thing in their favour — they are small. So, while they form a barrier to light waves, whose wavelengths are roughly the same size as the dust grains, they are practically transparent to radiation of longer wavelengths. An astronomer 'tuning in' to the Galaxy at infra-red or radio wavelengths will not 'see' any dust grains at all, and can look straight across the Galaxy.

But, even so, our position gives us only a worm's eye view of our star city. Radio astronomers still have to face the complicated task of disentangling dozens of signals piled on top of each other along the same line of sight, and that is one reason why the exact shape of the Galaxy is open to debate even now. At least nature has co-operated by providing a strong clear signal to help in the job of mapping. It comes from the cold hydrogen gas that pervades the space between the stars.

About one-tenth of the detectable mass of our Galaxy (ignoring the dark matter) is in the form of interstellar gas, mainly hydrogen. Although the hydrogen is invisible to our eyes, the signal that it gives out at a precise wavelength of 21.1 cm is a veritable beacon to radio astronomers. Its real selling point as a mapping signal is that it is a single wavelength, not a broad band of wavelengths like visible light. Astronomers measuring its intensity and the slight Doppler shifts from the 21.1 cm rest wavelength (caused by the motions of gas clouds towards or away from us) can tell not only how much gas there is, but how far away it lies.

In the 1920s, the Dutch astronomer Jan Oort was one of the first to produce evidence that our Milky Way Galaxy is a spiral, by studying the motion of stars near the Sun. His pioneering spirit came to the fore again in the 1940s, when the Germans occupied The Netherlands and closed the observatories. Undaunted, Oort insisted on holding his astronomy group together by tackling purely theoretical problems. Oort realised that radio astronomy could extend his own researches into the structure of the Galaxy: an American researcher had found radio waves coming from space a few years earlier, but professional astronomers had not followed up this lead. What was needed, Oort argued, was a signal from gas in space that he could tune in to. He set this problem to a student, Hendrik van de Hulst, who calculated that hydrogen atoms would produce the 21 cm wavelength. In 1951, several radio astronomy groups around the world succeeded in picking up this radiation from the hydrogen in space, and so began to lay bare the spiral backbone of our Galaxy.

The study of hydrogen produced a broad-brush picture of our Galaxy – but astronomers more recently have found a better way to pin down the details of the spiral arms. The problem with hydrogen is simply that there is so much of it – we find it everywhere, between the arms (though more thinly) as well as in the arms. In the 1960s, radio astronomers discovered they could pick up radiation at particular wavelengths from other gases in space – not single atoms, this time, but molecules. The molecules can only survive in places where they are safely hidden from disrupting ultra-violet radiation – in the centres of the black clouds in space. And we know that dark clouds are strung like beads along the spiral arms of galaxies. So astronomers have now tuned into a convenient signal, from carbon monoxide at a wavelength of 2.6 mm, to produce the first detailed maps of our Galaxy, not just in our vicinity, but right across to the far side of the Milky Way.

By observing radio waves from nearby dark clouds, astronomers have been able to find out just what is going on in their murky depths. First, they have detected signals from an astonishing number of different molecules – a rich and varied collection consisting not just of two or three atoms, but up to thirteen atoms strung together. To date, over sixty varieties have been discovered.

Some are very simple, like carbon monoxide (CO), water (H_2O) and cyanogen (C_2N_2). Others are decidedly complex, such as methyl cyanoacetylene (CH_3C_2CN) and cyanohex-atri-yne ($HC_2C_2C_2CN$). There is even ethyl alcohol (C_2H_5OH) – in one cloud complex alone there is enough to fill the Earth with neat whisky! To understand what is going on, astronomers are working closely with chemists – although the interstellar chemistry

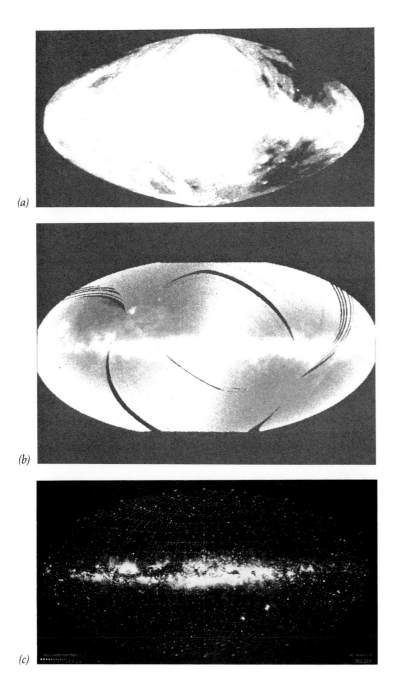

5. *The Milky Way in different wavelengths.* (a) *73 cm long radio wavelengths. Radio emission is most intense in the galactic plane but loops extend beyond this region.* (b) *60–100 μm infra-red wavelengths. This view, produced from data collected by IRAS, is dominated by emission from heated dust grains. The dust grains are in regions where stars are being formed.* (c) *Optical wavelengths. A painting of the whole sky by M. and T. Keskula of the Lund Observatory. The optical images of nebulae and the observed brightness of stars down to about 10th magnitude are shown. The image is convincing evidence that we live in a disc-shaped galaxy.*

laboratory is rather different from what we are used to on Earth. A research chemist can often produce a reaction only by heating the substances, to encourage the molecules to swap their atoms. But, in the interstellar clouds, temperatures are not far above absolute zero. Reactions must proceed extraordinarily slowly – but, unlike the laboratory chemist, nature has millions of years in which to perform these experiments.

The dark 'molecular clouds' are the maternity wards of the Galaxy: in the centre, gas is tightly compressed into clumps, which then pull themselves together by their own gravity. In the process, the clumps become hot, and shine brightly at infra-red wavelengths. These clumps are not yet stars, for they have not switched on the central nuclear furnaces that give a star its permanent source of energy. Astronomers call them 'protostars'. Like radio waves, infra-red can penetrate the surrounding dust, and this radiation should provide a sure guide to the presence of protostars – the 'Holy Grail' for astronomers studying starbirth.

Unfortunately, even though this radiation can travel through light-years of dark dust, and then through thousands of light-years of interstellar space, it runs into an obstacle when it meets the thin layer of the Earth's atmosphere. Water vapour and carbon dioxide in the air readily absorb the radiation. Only by working on chilly mountain tops – or, better, with telescopes in space – can astronomers study the infra-red sky. A real breakthrough came in 1983, with the Infra-red Astronomical Satellite (IRAS) – a refrigerated telescope in orbit around the Earth. IRAS surveyed the whole sky, and picked out half a million sources of infra-red.

Astronomers interested in starbirth looked eagerly through IRAS's results to find sources of infra-red that lay deep in dark clouds. There they picked out the first protostars. Further study of these is providing clues to the way in which our own planetary system was formed, and the first results indicate that planets are a natural by-product of the birth of stars, so that worlds like the Earth may well be common in the Galaxy.

Although the details of starbirth are now becoming clear, astronomers still are not sure how the interstellar material becomes compressed into molecular clouds in the first place. One possible agent is the shock from a supernova. We can see the effects of supernovae in the general interstellar medium, which has a structure rather like a Gruyère cheese. It is made up of great arcs and loops of gas that are punctuated by low-density bubbles – regions swept clean by ancient supernova explosions. Around the edges of the bubbles are denser clouds – some sufficiently compressed that stars could form within.

Supernovae can trigger the birth of stars in their own vicinity; but there is a Galaxy-wide process at work as well. According to the 'density wave theory', the beautiful pattern of the spiral arms marks a huge set of shock waves travelling around the Galaxy. Stars do not reside permanently 'in an arm' or 'between arms': as each star travels around the centre of the Galaxy in its own orbit (once every 250 million years, in the Sun's case), it passes in and out of the spiral arms.

The stars making up any particular spiral arm are, therefore, constantly changing. Although the constant interchange seems to suggest that the arm will quickly lose its identity, gravity ensures that this is not the case. The very existence of the arm means that gravity is stronger here – because of the bunching of the stars – than outside. As the stars move around the Galaxy, they are actually pulled quickly towards an arm, and

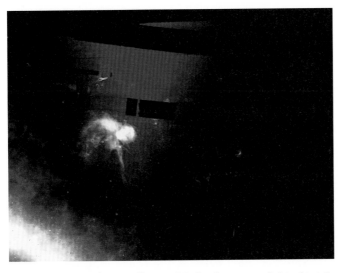

6. *A star-forming region in the constellation of Ophiuchus is revealed in this infra-red image by IRAS.*

7. *Artist's impression showing the spiral structure of our Galaxy. The Sun lies in a minor spiral arm between two more prominent spiral arms.*

then, once inside, are slowed down. Getting out of an arm is a slow business, too, against the force of the arm's extra gravity. Then it is a glide between arms before being accelerated on to the next. The net result is that a star spends longer in a spiral arm, and so at any given time more stars live *in* the arms than between them. The arms are density fluctuations in the Galaxy's disc where the gravity is higher. Computer calculations show that once this perturbation is set up, it will slowly propagate around the Galaxy at its own pace, preserving its spiral shape as the stars drift through the pattern.

Although the stars – as the main component of the disc – set up the spiral pattern, they are not as affected by it as is the gas in space. When the gas from the region between the arms enters an arm, it cannons into the gas already there – setting up a shock wave that compresses the gas (and the dust it contains) into the dense molecular clouds that are the sites of starbirth. That is why we find regions of starbirth strung along the spiral arms; and why astronomers can, conversely, use these regions to map out the spiral arms of the Galaxy.

There is one final region of our Galaxy that we are only just beginning to understand: the very centre. Like other spiral galaxies, the Milky Way has a large central bulge of stars that is some 25 000 light-years in diameter and 10 000 light-years thick. It is composed of old red stars which formed long ago, shortly after the collapse of the large gas cloud that formed the Galaxy. Unlike the surrounding disc, the bulge has virtually no dust and gas. Until quite recently, astronomers considered it to be an inert, uninteresting place, and its innermost regions could anyway never be seen: the dust grains in our Galaxy's disc produce more than thirty magnitudes of absorption along the way – that means that the centre is dimmed many thousands of millions of times.

Mapping the spiral structure of our Galaxy with the 21 cm hydrogen radiation back in 1957, Jan Oort was able to penetrate closer to its centre than ever before. He noticed something strange. Instead of spinning uniformly around the centre, some of the spiral arms he could detect appeared to be moving outwards – away from the centre. Intrigued, he plunged on inwards. He found several more gas clouds, some of them fragments of arms, even jets, which all shared this outward motion.

As the resolution of radio telescopes improved in the 1960s, other astronomers reported peculiar features close to the centre of the Galaxy. Surrounding the very centre, and 1500 light-years across, there is a ring of dark clouds, rich in molecules: it is expanding at a rate of half a million kilometres per hour. Embedded in this ring are huge star-forming clouds, each ten times the size of the Orion Nebula. Inside the ring – in a part of the Galaxy not famed for its abundance of interstellar gas – is an extended region of hot hydrogen centred on the Galaxy's heart.

There are stars here in mind-boggling quantities. Although we can never see them with ordinary telescopes, infra-red astronomers, slicing through the intervening dust, report that the stars appear to be forming within a light-year of the galactic centre. Surrounding the centre is a dense cluster of a million ancient red and yellow stars, so tightly packed that the distance between neighbours dwindles from the four light-years we find in the neighbourhood of the Sun to just four light-*days*.

In the very centre of all this, there lurks a source of radio waves that is powerful but extremely small – less than the span of Jupiter's orbit about the Sun. Many astronomers

8. *The galactic centre imaged by the Anglo-Australian Telescope.*

believe that this core is dominated by a massive black hole, some three million times more massive than our Sun, which was created early in our Galaxy's history. Black holes are potent engines. Gas and stars that come too close are snatched, torn, swirled, heated, accelerated. The debris forms an accretion disc – a whirlpool of hot gases – that glares fiercely before its matter disappears into the black hole for ever. It is this glare that radio astronomers may be picking up from the direction of our galactic centre.

The power of an accretion disc can show itself in other ways, for example as shock waves that could drive away the surrounding gas clouds at explosive speeds. Perhaps these shocks have caused the formation of the young stars near the Galaxy's heart, and they could be responsible for the outward moving gas streams further out. By backtracking the motion of this gas, some astronomers claim that the accretion disc has had violent spasms of activity several times in the past twenty million years, presumably when a large lump of gas – perhaps a sacrificial star – has fallen into the accretion disc.

Not everyone agrees. Other astronomers point out starbirth itself is a violent process. IRAS discovered some galaxies undergoing an immense hiccup of starbirth – so intense that their output of infra-red radiation is one hundred times more powerful than their light. Although the Milky Way is not a 'starburst galaxy' of this magnitude, perhaps a mini-starburst at its centre is responsible for most of the strange events there.

Only further research will show which idea is correct. But it is fascinating, and not a little exciting, to realise that our Galaxy, for so long regarded as a serene unchanging whirlpool of stars, has a violent secret hidden in its heart.

13

Active galaxies and quasars*

MALCOLM LONGAIR

The discovery of galaxies with active galactic nuclei

Galaxies are the building blocks which define the large-scale structure of the universe. Most of their visible mass is in the form of stars, and it is the gravitational pull of the stars on one another which holds the galaxy together, although there may also be dark matter present in the outer regions of giant galaxies. Many different types of galaxies have been identified – normal spirals, barred spirals, lenticulars or SOs, ellipticals and irregulars, as well as a wide variety of intermediate types.

Until the 1950s, that was all there was to extragalactic astronomy. About that time, however, it was realised that there exist various types of galaxy in which all the activity could not be attributed to the processes by which stars normally emit energy. These galaxies can be referred to collectively as *galaxies with active galactic nuclei*. Historically, the first class of galaxies with active nuclei to be discovered were the *Seyfert galaxies*, in 1942. These appear to be spiral galaxies but possess star-like nuclei (Figures 1a and b). When the spectra of these galaxies were studied, it was found that the emission lines are very broad and strong, unlike those found in any of the emission line regions in normal galaxies. Many Seyfert galaxies have now been found, and they are among the most important classes of active nuclei because they are relatively common.

---

*This contribution is an abbreviated version of a Section of my review 'The new astrophysics', which appeared as Chapter 6 of *The New Physics* (ed. P. C. W. Davies), Cambridge University Press, 1989. More details of the arguments, and much more of the astronomical and astrophysical context, are described in that review.

Malcolm Longair

The next class of active galaxy to be discovered were the *radio galaxies*, which are very powerful sources of radio waves. By the mid-1950s, it was established that these galaxies must be sources of vast amounts of high-energy particles and magnetic fields. A few of these galaxies had star-like nuclei and were called *N-galaxies*. They had properties in

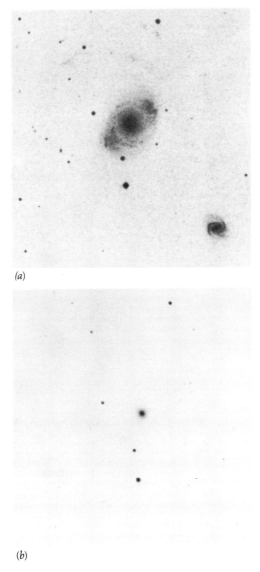

(a)

(b)

1. Two images of the nearby Seyfert galaxy NGC 4151. This was one of the galaxies noted by Karl Seyfert in his first catalogue of galaxies with star-like nuclei and broad emission lines. (a) An image of NGC 4151 in a long exposure showing the inner structure of the disc of the galaxy. In an even longer exposure, spiral arms are clearly visible in its outer regions. (b) A short exposure image of NGC 4151 showing the star-like nucleus of the galaxy.

179

common with the Seyfert galaxies and also had strong broad emission lines in their spectra but the relation between these phenomena was not clear at that time.

In the early 1960s, the first *quasars* were discovered. Among the objects associated with bright radio sources were a few 'stars' which had quite unintelligible optical spectra. The radio source 3C 273 (Figure 2) was one of those which had been found in early radio surveys of the sky. 3C 273 looked exactly like a star on a photographic plate and was found to be variable in brightness over a period of years. The remarkable discovery made by Maarten Schmidt in 1962 was that 3C 273 turned out to lie at a distance similar to that of the most distant galaxies whose distances were readily measurable at that time. The object was called a *quasi-stellar object* because it looked like a star but clearly could not be any normal sort of star at this very great distance. 3C 273 is more than 1000 times more luminous than a galaxy such as our own and yet this luminosity varies on remarkably short time scales. Following this great discovery, many more quasars were found, all of them characterised by stellar appearance and very great distances. In addition to those which are strong radio sources, *radio-quiet quasars* were discovered in 1965 which were

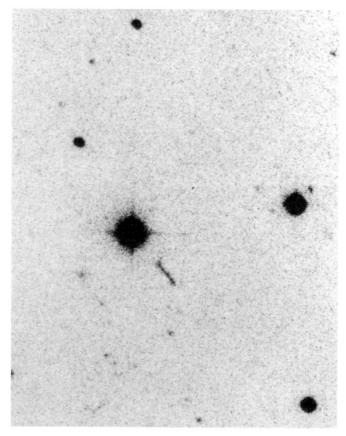

2. *The quasar 3C 273. This deep image taken by Halton Arp shows the quasar and its famous jet pointing towards the bottom right-hand corner of the image. The faint smudges to the south of the quasar are galaxies at the same distance as the quasar.*

3. *This image was obtained by the 2.2 m telescope at the European Southern Observatory. It is of object PHL 1222, in the constellation of Pisces, which has been resolved into the closest pair of quasars ever observed.*

just as remarkable optically as the radio quasars but which were not powerful sources of radio emission. The quasars are among the most extreme examples of active galactic nuclei known. That they actually are the nuclei of galaxies has been convincingly demonstrated by the fact that the underlying galaxies in some of the more nearby examples have been detected.

Among the most extreme examples of active galactic nuclei are the objects known as *BL-Lacertae* or *BL-Lac* objects, which were discovered in 1968. These are similar to the radio quasars but they differ from them in two respects. First, they vary extremely rapidly in luminosity, variations being detected on time scales of days or less; they must therefore be very compact indeed. Secondly, their optical spectra are normally featureless and the continuum radiation is strongly polarised. It is plausible that, in the BL-Lac objects, we observe more or less directly the primary source of energy in active nuclei.

It should be emphasised that the most active of these systems are very rare indeed. It is because they are so luminous and have such distinctive properties that they can be discovered relatively easily. The astrophysical problem is to understand how they can generate such vast amounts of energy from very compact volumes. The current consensus of opinion is that these different types of active nuclei are all basically the same. There exists some ultra-compact source of energy in these objects, and precisely what one observes depends upon the environment of the compact object and the way in which it obtains its fuel. Even rather quiescent nuclei like the centre of our own Galaxy possess small-scale versions of the phenomena we see in quasars and other active nuclei. It is likely that there is a continuity in activity which runs from systems such as the centre of our own Galaxy through the Seyfert galaxies and N-galaxies to the quasars and BL-Lac objects.

The basic physics of active galactic nuclei

The understanding of the extreme properties of the most active galactic nuclei almost certainly requires new physics in physical conditions quite different from those encountered elsewhere in the universe. The key pieces of observational data are their extreme luminosities and, even more important, the fact that this emission varies rapidly with time. The most rapid variations are found in some of the BL-Lac objects and in the X-ray variability of some Seyfert galaxies (Figure 4). Significant variations on the time scale of about one hour have been found in sources which lie at cosmological distances.

Let us show how these observations lead inevitably to the conclusion that there must be supermassive black holes in active galactic nuclei and quasars. First of all, let us review the properties of black holes. All stars are held up by some form of pressure which prevents them collapsing under the attractive force of their own self-gravity. Stars like the Sun are held up by the thermal pressure of hot gas, whereas white dwarfs and neutron stars are held up by purely quantum mechanical degeneracy forces. The neutron stars, which typically have masses approximating that of the Sun but have radii only about 10 km, are the last known forms of stable star. For more compact objects, the attractive gravitational force of gravity becomes so strong that, according to classical physics, no force can prevent collapse to a physical singularity. These singularities in the fabric of space-time are what are known as black holes. To summarise their properties very briefly, a spherically symmetric black hole of mass M possesses a characteristic radius known as the *Schwarzschild radius* $r_g = 2GM/c^2$, which represents the effective radius of the black hole. Putting in the values of the constants, we find that $r_g = 3(M/M_\odot)$ km so that solar mass black holes have $r_g = 3$ km. The key points about the black holes are as follows:

(i) Radiation cannot escape from within the radius r_g because the force of gravity is too strong. This is why the black hole is called 'black'.

(ii) Matter falling within the Schwarzschild radius cannot escape outside r_g. This is why it is called a 'hole". Collapse to the central singularity takes place in a finite amount of proper time for the infalling object, but to the external observer the collapse takes an infinite amount of time to reach r_g. The reason for this difference is that the gravitational red-shift of radiation emitted from the radius r_g becomes infinite, although there is no real physical singularity at this radius.

There are in addition rotating black holes, often known as *Kerr black holes*. They are somewhat more complicated than the spherically symmetric case because the surfaces of infinite red-shift and the radius within which collapse to the singularity is inevitable do not coincide.

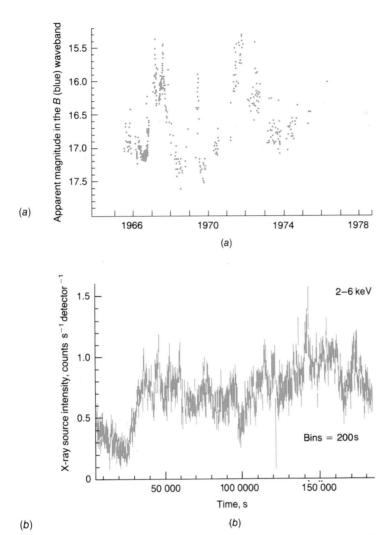

4. Examples of the variability of the emission of active nuclei. (a) The optical variability of the radio quasar 3C 345 in the B waveband. (b) The X-ray variability of the Seyfert galaxy M C G-6-30-15. Strong and continuous variability can be seen in the X-ray variability down to time scales of several hundred seconds.

One key aspect of the astrophysics of black holes is that there is a last stable circular orbit about the black hole such that a particle can remain in a stable circular orbit if its radius is greater than a critical value but, if its radius is less, the particle inevitably spirals into the black hole. If the black hole is rotating as fast as is allowed, i.e. (roughly speaking) its gravitational energy is the same as its rotational energy, the last stable circular orbit lies at a radius of only $0.54r_g$, whereas in the spherically symmetric case the last stable orbit lies at $3r_g$. This means that it is possible to tap much more of the energy of matter falling into a rotating black hole than onto the spherically symmetric variety. Detailed calculations show that, in the spherically symmetric case, about 6% of the rest mass energy of the infalling matter can be released, whereas in the case of a maximally rotating black hole this value increases to about 43%. Thus, black holes are potentially extraordinarily powerful sources of energy for high-energy astrophysical phenomena.

Let us now look at the relevence of these ideas to active galactic nuclei. We will start by combining two simple but crucial results which come from elementary astrophysics. The first of these is the expression for maximum luminosity which any object of mass M can have. The physics of this argument is simply that, if the object is too luminous, the radiation pressure acting on the material of the object will blow it apart. In other words, an object of a given mass must be able to hold itself together, despite the fact that it is releasing enormous amounts of energy. This maximum luminosity is known as the *Eddington limiting luminosity* and can be written as

$$L_{Edd} = 1.3 \times 10^{31}(M/M_\odot)W \qquad (1)$$

This luminosity is expressed in watts, and the mass of the object is expressed in units of the mass of the Sun, $M_\odot$. This sets a useful upper bound to the luminosity of an object of mass M. Notice that the result is independent of its radius.

The second result is the *causality* relation, which states that variations in the intensity of a source of size r cannot be observed to take place on time scales less than r/c since it takes this time for light and electromagnetic waves to travel from one side of the source to the other. The smallest size which a source of mass M can possibly have is roughly the Schwarzschild radius of an object of mass, and hence the lower limit to the time scale of variations is the Schwarzschild radius divided by the velocity of light

$$T \gtrsim r_g/c \approx 10^{-5}(M/M_\odot)\,s \qquad (2)$$

Thus, according to relation (1), a black hole 10^8 times the mass of the Sun can have luminosity up to $\approx 10^{39}\,W$ and, from (2), variations in this luminosity could occur on time scales as short as roughly half an hour. Thus, supermassive black holes have all the necessary properties for a successful theory of the most active galactic nuclei. To complete the picture we recall that the efficiency of energy release by accretion onto black holes is very high. In addition, once a black hole has been spun up by the matter falling into it, the rotational energy of the black hole can be tapped subsequently if the rotation is coupled to the surrounding interstellar medium by a magnetic field.

Direct observational evidence for compact masses of this order in the nuclei of Seyfert galaxies has come from long-term monitoring of their intensities and spectra. The most extreme types of Seyfert galaxies have very broad, strong emission lines which are variable in intensity, and also strong variable continuum radiation. The breadth of the lines is attributed to mass motions of gas clouds close to the nucleus itself. One remarkable result, which has been obtained by a consortium of European astronomers using the International Ultraviolet Explorer, has been that there is a time delay between the variations observed in the continuum spectrum and the corresponding variations seen in the broad line spectrum in the Seyfert galaxy N G C 4151. This indicates convincingly that the broad

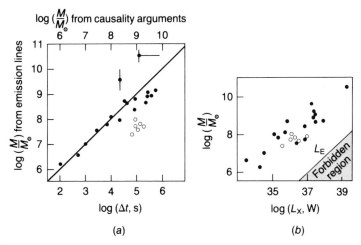

5. (a) *Comparison of the mass estimates of active galactic nuclei from the variability of their X-ray emission and from dynamical estimates by Wandel and Mushotzky. The solid dots are quasars and Type I Seyfert galaxies, and the open circles are Type II Seyfert galaxies.* (b) *Comparison of the inferred masses and luminosities with the Eddington limiting luminosity, $L_{Edd} = 1.3 \times 10^{31} (M/M_\odot)$ W. It can be seen that all of these objects lie well below the Eddington limit.*

line regions are photoionised by the continuum radiation from the nucleus and also enables a direct estimate of the distance of the clouds from the nucleus to be made since the ultraviolet photons travel from the nucleus to the clouds at the velocity of light. The combination of this distance with the velocities of the clouds enables the mass of the central region to be determined, and this turns out to be about $10^9 M_\odot$. Arguments of a similar type have been used by Amri Wandel and Richard Mushotzky to show that, for those Seyfert galaxies which are variable X-ray sources, masses determined from the causality relation (2) and dynamical masses derived from the line widths and inferred distances of the broad emission line regions from the nucleus, are in remarkable agreement over a considerable range of black hole masses, $10^6 \leqslant M/M_\odot \leqslant 10^{10}$ (Figure 5a). Typically, the luminosities amount to about 1 to 10% of the Eddington limit (Figure 5b).

The prime ingredients of active galactic nuclei

It is convenient to divide the necessary ingredients of active galactic nuclei into two types: the *primary ingredients*, which probably have to be produced in the vicinity of the black hole itself, and *secondary phenomena*, which result from the interaction of these primary ingredients with the environment of the black hole. Figure 6 shows a schematic diagram of some of the components of any successful model. The primary ingredients are the *intense non-thermal continuum radiation* and *fluxes of high-energy particles*. The secondary phenomena arise from the interaction of these components with the surrounding medium, in particular gas clouds close to the nucleus and the ambient interstellar and intergalactic gas. The former gives rise to the strong emission line spectrum observed at optical, ultra-

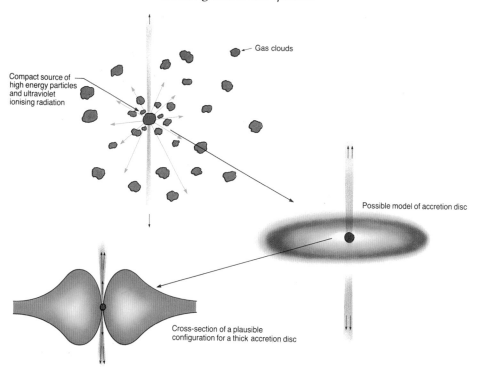

Gas clouds

Compact source of
high energy particles
and ultraviolet
ionising radiation

Possible model of accretion disc

Cross-section of a plausible
configuration for a thick accretion disc

6. A schematic diagram showing the necessary ingredients of a model of an active galactic nucleus. There must be a compact source of high energy particles and intense ultra-violet ionising radiation in the very centre. These may be produced close to the central engine, which is likely to be a super massive black hole. The ultra-violet radiation may be the emission of the relativistic particles accelerated close to the central engine or the thermal emission from the inner regions of the accretion disc. The nucleus is surrounded by gas clouds, which are heated and excited by the ionising radiation from the nucleus. The clouds closest to the centre have high particle densities, values of about $10^{16}\,m^{-3}$ being necessary to de-excite the forbidden line radiation from these regions. These are the origin of the broad-line emission seen in quasars, Type I Seyferts and broad line radio galaxies. Further out, the clouds are less dense, $\sim 10^{10}$–$10^{12}\,m^{-3}$, and these are responsible for the narrow-line regions observed in Type II Seyfert galaxies and some of the narrow-line radio galaxies. It is inferred that there must be some mechanism responsible for the collimation of the beams of high energy particles. The inserts to the diagram show a possible structure for the accretion disc and a thick disc surrounding the black hole. These may be responsible for the collimation of the beams.

violet and infra-red wavelengths, whilst the interaction of beams of high-energy particles with the interstellar and intergalactic gas give rise to beams and jets in extended radio sources.

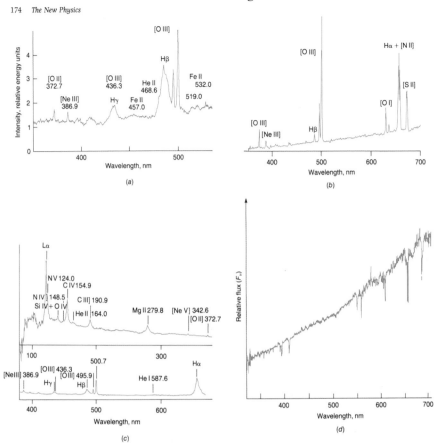

7. *Typical spectra of different classes of active galactic nuclei. (a) The Type I Seyfert galaxy III Z W 2 – the spectrum exhibits the characteristic broad emission lines of the resonance lines of hydrogen while the forbidden lines of oxygen and neon are narrow. (b) The Type II Seyfert galaxy Markarian 348 (Mk 348) – the lines are much narrower than in the case of Type I Seyfert galaxies and forbidden lines of the same line width as the permitted lines are observed. (c) A composite quasar spectrum extending from 100 to about 700 nm – the spectrum contains very broad emission lines with a strong nonthermal continuum. (d) The BL-Lac object PKS 0215 + 015 – there are no spectral lines observed in the spectrum which can be used to measure the red-shift of the object. Narrow absorption line systems are observed at red-shifts of 1.3449, 1.5494 and 1.6494. These are presumably foreground absorbers and hence the BL-Lac object must lie at a larger unknown red-shift. The continuum spectrum can be represented by a power-law spectrum of the form $I_\nu \propto \nu^{-2.11}$.*

The continuum spectrum

Figures 7(a)–(d) show typical optical spectra of Types I and II Seyfert galaxies, a quasar and a BL-Lac object. In all cases most of the luminosity is in the continuum spectrum rather than in the lines; furthermore, the variability of the source is almost entirely

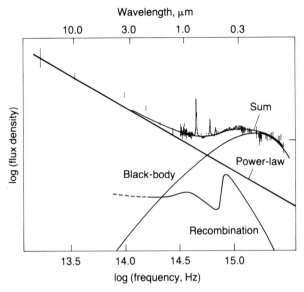

8. *The optical-ultra-violet spectrum of the quasar 3C 273. The continuum has been decomposed into a 'power-law' component, a component associated with recombination radiation and a 'blue bump' component, which has been represented by a black-body curve. The prominent Balmer series in the optical waveband which led to the discovery of the high red-shifts of quasars can be seen.*

associated with the continuum spectrum. The continuum spectrum is unlike any stellar or galaxy spectrum, the latter being the integrated light of many different types of star. Very often the continuum spectrum can be represented by a power-law within the optical and infra-red wavebands, most objects having spectra which can be described by $I_v \propto v^{-\alpha}$, with the average value of α close to unity but with considerable dispersion about this value. For a number of bright objects, it has been possible to undertake more detailed decompositions of the continuum spectrum in the optical and ultra-violet wavebands, and these show evidence for an excess at ultra-violet wavelengths as compared with what is expected from extrapolation of a power-law spectrum. The excess is probably similar in character to the 'blue bump' seen in the ultra-violet spectrum of 3C 273 (Figure 8).

Many quasars and active nuclei have been detected by X-ray satellites such as the Einstein X-ray Observatory and Exosat, and it appears that their X-ray spectra form a natural extension of the optical spectrum into the X-ray region. The optical-to-X-ray spectra of most quasars can be described by a power-law with spectral index $\alpha \approx 1$, this form of spectrum indicating that roughly as much energy is radiated per decade of frequency in the optical and X-ray wavebands.

The far ultra-violet region of the spectrum is important for understanding the excitation of the gas clouds which produce the strong emission lines surrounding active galactic nuclei. For the most extreme quasars, a convincing case can be made that the strong line spectra result from photoionisation and excitation of the gas clouds surrounding the nucleus by the continuum radiation from the nucleus (Figure 6). One of the attractions of the photoexcitation picture is that a power-law continuum spectrum, containing photons with a very wide range of energies, can account for the observation of a very wide range of different ionisation states of elements in a single spectrum.

The continuum radiation is also known to be polarised. The strongest polarisation is found in the cases of the BL-Lac objects in which up to 30–40% linear polarisation has been observed. In a few important cases, polarisation changes have been followed through strong radio and infra-red outbursts, and these prove important in understanding the magnetic field geometry of the source regions. It appears that the radiation associated with the 'blue bump' is not polarised since the total percentage polarisation decreases when it contributes significantly to the total continuum intensity. This is consistent with the blue bump being mostly thermal radiation.

High-energy particles

It is likely that very high-energy particles are accelerated close to the nucleus itself. The characteristics of a power-law continuum spectrum, polarised emission and rapid variability are typical of those expected of the emission of ultra-relativistic electrons. The most direct evidence for the presence of ultra-relativistic electrons in the nuclei of active galaxies comes from very long baseline interferometric (VLBI) studies of radio quasars and BL-Lac objects at centimetre wavelengths. This radio astronomical technique enables very fine-scale angular structure in radio sources to be measured, the longest baselines between radio telescopes now being limited by the diameter of the Earth. Angular resolution of a fraction of a milliarcsecond have now been achieved. Combining the angular sizes of these ultra-compact radio sources with their flux densities, the *brightness temperature* of the source region can be determined, the highest values found being about 10^{11} K. It is interesting to compare this figure with the temperature of electrons which have kinetic energies equal to their rest mass energies, $kT \approx m_e c^2$, i.e. $T \approx 5 \times 10^9$ K. Assuming the emission is the radiation of high-energy electrons, this proves that they must have ultra-relativistic energies, i.e. $E > m_e c^2$ since particles cannot radiate at brightness temperatures greater than their thermodynamic temperatures. The angular sizes which are currently accessible for distant quasars by VLBI techniques correspond to physical sizes which are still much greater than the Schwarzschild radius of a $10^8 M_\odot$ black hole. The smallest physical scales which have been studied by VLBI correspond to about one parsec (1 pc), whereas the Schwarzschild radius of a $10^8 M_\odot$ black hole is only 10^{-5} pc. It seems entirely plausible, however, that the relativistic matter is accelerated close to the black hole because the continuum emission observed in the optical, infra-red and ultra-violet wavebands also has all the characteristics of the emission of relativistic electrons.

The most likely mechanism for the emission of radiation by relativistic electrons in active galactic nuclei is the synchrotron radiation process, in which they move in spiral trajectories through the magnetic field in the source. This process has the characteristic feature that, because of the very strong beaming of the radiation associated with the relativistic motion of the electron, the radiation spectrum emitted by each particle is intrinsically broad band. As a result, a power-law distribution of electron energies is guaranteed to produce a power-law emission spectrum. The radiation is linearly polarised in an aligned magnetic field with a maximum percentage linear polarisation of about 70%. The intrinsic polarisation angle of the radiation is perpendicular to the magnetic field direction for an optically thin source so that information about the magnetic field geometry can be obtained from these studies. It can be seen that these properties of synchrotron radiation match satisfactorily the observed properties of the optical, infra-red and radio continuum emission.

One of the most important facts derived from radio VLBI observations is that the radio jets observed on a large scale in radio galaxies have their origin in the nuclei themselves. The large-scale jets can be traced back more or less continuously to jet structures on a scale of about 10 pc or less, many of which show evidence that the radio components are moving away from the nucleus at velocities apparently in excess of the speed of light. We will return to this intriguing topic later, but for present purposes this observation gives us two further pieces of evidence. First, it is clear that *something* relativistic is escaping from the nuclear regions. Second, the observation of an axis of ejection of relativistic material strongly suggests that there is some axisymmetric structure such as an accretion disc or rotating black hole in the nucleus which provides a natural axis for the production of collimated beams of particles.

The black hole

When matter falls towards a black hole under gravity, it is most unlikely that it falls straight into the hole but rather that it will take up a bound orbit about the hole. The infalling matter soon settles down to what is known as an *accretion disc* about the hole in which the matter slowly drifts in towards the last stable orbit. To do this, energy has to be dissipated, and it is this frictional energy released by matter as it reaches its last stable orbit which is probably the main source of energy generation in black holes. In the case of active galactic nuclei and quasars, the black holes must be very massive, and the radiation flux from the accretion disc is so intense that it strongly influences the structure of the disc. In fact, the problem of finding self-consistent solutions for accretion discs about supermassive black holes has not been solved. A possible configuration which has many desirable features is the thick disc model shown in Figure 6. The thick disc develops two oppositely directed funnels along the axis of the accretion disc, and these may be important in collimating the beams of particles which are inferred to originate close to the nucleus. In pictures like this, it is not clear exactly what the origin of the various primary components are. Simple theory suggests that the temperature of the radiation scales is $M^{-1/4}$ and so if solar mass black holes produce X-ray emission, $10^8 M_\odot$ black holes may radiate in the ultra-violet region of the spectrum. One possibility is that the 'blue bump' seen in the spectrum of 3C 273 (Figure 8) and other active nuclei may be the thermal emission of the accretion disc about a supermassive black hole.

Besides the emission of the disc, there may well be processes associated with the black hole itself which can give rise to the acceleration of charged particles. Although an isolated black hole cannot possess a magnetic field, a magnetic field can be associated with black holes if it is anchored to the surrounding ionised plasma. A massive rotating black hole thus bears some resemblance to the magnetised rotating neutron stars which are certainly effective sources of relativistic electrons. Richard Lovelace, Roger Blandford and Roman Znajec, have shown how particles can be accelerated in the vicinity of rotating black holes, and there is every possibility that the continuum radiation may originate in the radiation of these particles by synchrotron, inverse Compton or curvature radiation.

The emission line regions

One of the major areas of study in active galactic nuclei is the nature of their emission line spectra which originate close to the central regions. There is evidence for a wide range of physical conditions in different classes of active nuclei, the most important parameters being the intensity and spectrum of the ionising radiation and the particle densities within the gas clouds.

The general picture which emerges is as follows (see Figure 6). The emission line regions which originate closest to the black hole are responsible for the broad line emission observed in the most active nuclei, the Seyfert I galaxies and the quasars (Figures 7a and (b). The most direct evidence for this is the fact that the emission profiles are found to be variable on a time scale of months and hence they must be very close to the nucleus. They must also be rather dense regions because there are no forbidden lines accompanying the strong broad emission lines. Thus, in the nuclear regions within about 0.1 pc or less of the black hole, there are dense clouds with a large velocity dispersion. It has not been possible to distinguish what these motions are, whether they represent turbulent, rotational, expansion or contraction velocities, although careful study of the time evolution of the line profiles during an outburst from the nucleus can, in principle, distinguish between these. Further out from the nucleus are the less dense clouds responsible for the narrow emission line spectrum seen in Seyfert galaxies (Figure 7b). These emit strong permitted and forbidden lines with similar line widths and also exhibit a wide range of ionisation states.

In many Seyfert nuclei, there is very often evidence for both types of gas cloud being present, the relative strengths of the broad and narrow line components varying from Type I Seyfert galaxies which are pure broad line systems to pure Type II Seyferts in which the narrow lines are dominant. Some of the radio galaxies which are very powerful radio sources have properties similar to Type I and Type II Seyfert galaxies, but there are important differences. Some of the radio galaxies have broad line spectra similar to the Type I Seyfert galaxies. Others have spectra more akin to the Type II Seyferts, but among the most distant and luminous radio galaxies there are several which possess very strong, narrow lines which do not originate in the nucleus. The emission line regions can extend to 50–100 kpc, greater than the size of the galaxy itself. It is these strong emission line radio galaxies which are the most distant galaxies yet observed, the largest red-shifts obtained so far being about 3.5. Finally there are even weaker nuclei in which only a low ionisation spectrum is observed. In these 'low ionisation, narrow emission line region' galaxies (LINERs), the excitation mechanism for the clouds could be either collisions or photoexcitation by ultra-violet radiation.

Radio emission from active galaxies and quasars

All galaxies are sources of radio emission for the same reason that our Galaxy emits radio waves – high-energy electrons are accelerated in supernova explosions and these are dispersed throughout the interstellar medium where they radiate radio waves by the synchrotron process. However, these are very weak radio emitters indeed compared with what we mean by the term *radio galaxy* or *radio quasar* in which the radio luminosity can exceed that of our own Galaxy by factors of 10^8 or more. The production of the high-energy particles is totally different in these powerful radio sources.

What is observed is that the nuclei of the radio galaxies and quasars are themselves strong radio sources, and jets or beams of relativistic material are ejected from them to form large-scale radio jets, double radio sources, and so on. The example of the brightest extragalactic radio source in the northern sky, Cygnus A, observed by the Very Large Array (VLA), is shown in Figure 9. The important properties of these sources are, first, their very large radio luminosities, implying the production of enormous fluxes of relativistic electrons and magnetic fields and, second, the fact that the energy generated in the nucleus of a galaxy can lead to the expulsion of relativistic gas in collimated jets far beyond the confines of the parent galaxy and into the intergalactic medium. The

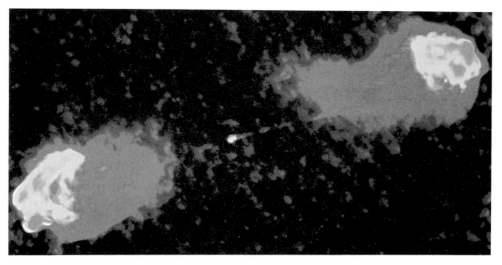

9. An example of the radio structure of the extended radio source Cygnus A as observed by the US Very Large Array (VLA). The radio source is associated with a giant elliptical galaxy which is the brightest member of a rich cluster. There is a compact radio source coincident with the nucleus of the galaxy, and a very narrow collimated jet can be seen extending from the nucleus to the extended radio lobe to the right of the image. VLBI observations show that even on the scale of 1 milliarcsec, the nucleus is elongated in the form of a jet in the direction of the long narrow jet seen in the figure. There are hot-spots within the lobes where particles are accelerated probably because of the interaction of the jet with the intergalactic medium. There is a great deal of structure in the lobes, probably associated with the escape of the particles from the hot-spots and the interaction of this relativistic gas with the surrounding medium.

largest radio structures known have physical diameters greater than 1 Mpc, the record being about 6 Mpc.

The facts that the radiation has a power-law spectrum, that the radiation is linearly polarised, and that these huge radio structures are observed in the very diffuse conditions of interstellar and intergalactic space, makes the identification of the radiation mechanism with synchrotron radiation wholly convincing. The theory of this process can be used to set lower limits to the amount of energy which must be present in the jets and radio lobes. The energy requirements are greatest for the largest and most diffuse sources and are $\geq 10^{54}$ J, corresponding to the conversion of at least 500 000 solar masses of matter into high-energy particles and magnetic field energy. Thus, there must exist efficient mechanisms for converting rest mass energy into these forms and then injecting them into the extended radio lobes over a period of ten to one hundred million years, the typical lifetime of extended radio sources.

The discovery of radio jets within the diffuse radio structure, for example in the VLA radio map of Cygnus A (Figure 9), is a key observation for understanding how these radio structures come about. A schematic model for the evolution of double radio sources is depicted in Figure 10. A beam or jet of particles is emitted from the nucleus of the radio galaxy which 'burns' its way out through the interstellar and intergalactic gas much like a laser beam. At the interface between the beam and the static intergalactic gas a shock region is set up, similar in shape to the magnetospheric cavity about the Earth's magnetic field. Particles are accelerated in the head of the beam and are left behind by the advancing jet. The relativistic gas left behind expands to form a large cavity of mixed relativistic plasma and magnetic fields which produces the large radio lobes observed in diffuse sources. When the beam switches off, the large diffuse lobes are left to dissipate their energy, either by synchrotron radiation or inverse Compton scattering of the microwave background radiation or, most likely, simply by adiabatic expansion.

The basic problem is therefore to understand the nature of the jets which originate close to the nucleus itself. Figure 11 shows VLBI images of the central regions of the radio source 3C 273 taken at various epochs between 1977 and 1980. The radio structure consists of a jet of discrete radio blobs which move away from the nucleus in the direction of the much larger radio and optical component which lies at a distance of about 50 000 pc (see Figure 2). This is convincing evidence that the large-scale radio lobes are powered by beams of particles ejected from the nucleus. The figure also illustrates one of the very surprising features of these compact radio sources. The source appears to have expanded in size by about twenty-five light-years in only three years. This is an example of the phenomenon known as *superluminal radio sources*. The beam appears to be expanding away from the nucleus at a velocity about eight times the speed of light, which at first sight appears to contradict causality. This phenomenon has been observed in a number of the brightest compact radio sources and so it is not an uncommon phenomenon.

These sources have excited the greatest interest among theoreticians who were not long in providing a large number of clever models to account for the observation of components which move at velocities apparently in excess of the velocity of light. One model which has attracted particular attention is the *relativistic ballistic model,* in which a jet of radio-emitting material is ejected from the nucleus at an angle close to the line of

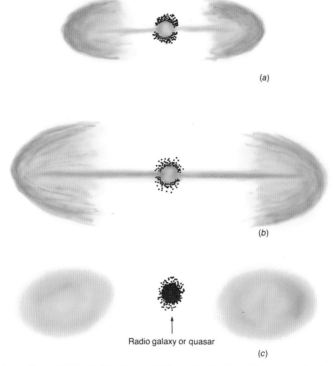

Radio galaxy or quasar

(c)

10. *A schematic model illustrating the evolution of a double radio source. The jets observed in VLBI observations are assumed to be associated with the continuous supply of energy in the form a of a relativistic beam of particles. (a) The beam burns its way through the intergalactic gas, and at the interface between the jet and the gas there is a contact discontinuity and a bow shock. Particles are accelerated in these regions which gives rise to the formation of 'hot-spots'. (b) The relativistic plasma expands producing an extensive region filled with relativistic particles and magnetic fields which form the extensive lobes seen in double radio sources. This configuration persists as long as the jet continues to supply energy to the outer lobes. (c) When the energy supply ceases, the diffuse lobes expand and decay in radio luminosity, principally because of adiabatic losses of the energy of the relativistic particles as the lobe expands.*

sight. If the jet moves at a velocity close to that of light, the jet almost catches up with its own radiation and it is observed to move away from the nucleus at a maximum velocity of γv, where $\gamma = (1 - v^2/c^2)^{-1/2}$ is the Lorentz factor and v is the velocity of the jet. The reason for this behaviour is illustrated in Figure 12. Thus, if v is close enough to c, the jet can appear to move arbitrarily rapidly without any violation of causality. An attraction of this picture is that, in addition to producing superluminal expansions, the radio luminosity of the advancing component is strongly boosted by the large blue-shift of the jet, and this can partly explain why many of the components appear to be one-sided. However, it is not at all clear that this model can explain everything. One problem is that there appear to be many such sources among the brightest radio sources and little

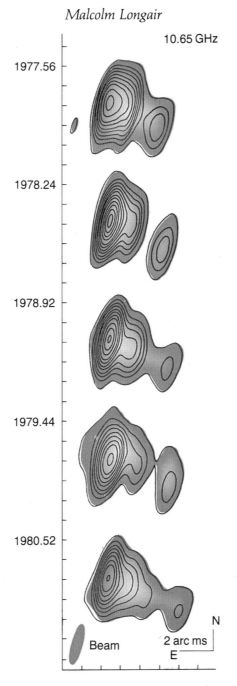

10.65 GHz

1977.56

1978.24

1978.92

1979.44

1980.52

Beam

2 arc ms

N

E

11. VLBI images of the nucleus of the radio quasar 3C 273 for the period 1977–80. The radio component is observed to move a distance of about thirty-five light-years in a period of four years, implying a superluminal expansion velocity of about 8–9c.

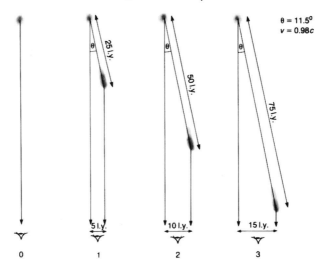

12. *Illustrating how a jet ejected at a relativistic velocity at an angle close to the line of sight can result in motion which appears to be superluminal to a distant observer. If the jet is ejected at an angle close to the line of sight at a velocity close to that of light, the jet almost catches up with its own radiation. The illustration shows the geometry and parameters necessary for observing an apparent expansion velocity of five times the velocity of light.*

room is left for the sources which are not pointing more or less towards us. The parent population from which we see only the small fraction of sources which are pointing more or less towards us is not at all obvious.

Although these observations give evidence for relativistic phenomena about 1 to 10 pc from the nucleus, this activity is still on a very much larger scale than that of the thick accretion discs and rotating black holes which may be the ultimate source of energy. Indirect evidence for relativistic streaming on the scale of light-days comes from analyses of the polarisation properties of the optical and infra-red continuum of BL-Lac objects. It is difficult to understand why the radiation is polarised at all because depolarisation of the radiation by the electrons responsible for the continuum emission itself by internal Faraday rotation is expected to be important in the extreme conditions of active galactic nuclei. The observation of polarisation and the rapid swings in the polarisation vectors of the continuum emission can be understood if the radiation is strongly beamed towards the observer by the relativistic bulk motion of the emitting electrons.

It is the combination of these separate pieces of evidence which strongly suggest that the beams of relativistic material originate very close to the accretion disc and the black hole in the active nucleus. It is inferred that this type of activity must continue over rather long periods because the extended radio components decay in radio luminosity by adiabatic expansion unless they are kept replenished by relativistic gas from the nucleus. The continued ejection of the relativistic gas in the same direction is guaranteed by the

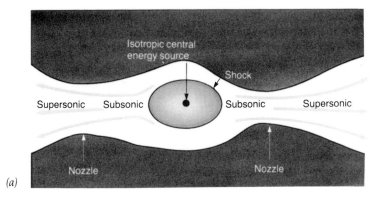

13. Two possible mechanisms for producing collimated jets of particles from the central regions of quasars and radio galaxies. (a) A gas-dynamical model in which a strong source of energy is located in a gaseous disc. The gas escapes from the region along the minor axis of the disc. In the process it may set up the type of 'nozzle' structure depicted in the figure. There is a shock surrounding the energy source. The flow of gas is subsonic until it passes through the nozzles beyond which the flow becomes supersonic as in a de Laval nozzle. (b) A 'flaring' disc model in which there is a powerful stream of relativistic gas ejected from the accretion disc about the black hole. Instabilities in the disc and the presence of a strong magnetic field lead to 'flares' on the surfaces of the disc which eject relativistic material in the direction perpendicular to the disc. The resultant jet of relativistic gas escapes at the relativistic sound speed $c/\sqrt{3}$.

fact that the rotating black hole will maintain its axis in the same direction over very long periods.

We still have not answered the question of the physical origin of the relativistic jets in active galactic nuclei. The schematic model shown in Figure 6 for a thick accretion disc with funnels along the axis of rotation may have something to do with the collimation mechanism, although it has yet to be established that these are stable structures. Two other possibilities are illustrated in Figure 13. In one case, the production of the jet of particles is attributed to electromagnetic processes occurring close to the rotating black hole, the rotation axis providing a natural axis along which the beams of particles are ejected. In the second model, the jet could be created by gas dynamical processes occurring in the dense regions surrounding the nucleus. The continuous generation of energy leads to the formation of a region of very high-energy density of high-energy particles. This relativistic gas will burst through the surrounding cloud in the direction of the poles of the cloud and form a collimated jet of relativistic gas. This model, due to Roger Blandford and Martin Rees, provides a possible link between the properties of the radio jets and the regions of high density gas inferred to be present in the broad line regions. It remains to be established how relevant these ideas are to the understanding of the origin of jets in active galactic nuclei.

The future

This brief summary has highlighted some of the key astrophysical problems which have come out of studies of active galaxies and quasars. These problems have opened up quite new vistas for astronomy and for fundamental physical processes. Besides the many challenges of understanding the astrophysics of gas, magnetic fields and high-energy particles in the extreme conditions of active galactic nuclei, there is the real possibility of studying directly the behaviour of matter in very strong gravitational fields. There is nowhere else in the universe where these studies can be undertaken.

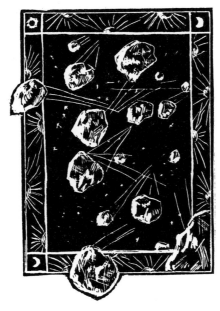

14

The first one second of the universe

PAUL DAVIES

It is the job of the astronomer to study the various objects which make up the universe. These include the Sun and planets, the many different types of stars, the galaxies, and the interstellar material. By contrast, the cosmologist is less concerned with the specific cosmic furnishings, more with the overall architecture of the universe. By 'the universe' cosmologists mean everything: the entire physical world of space, time and matter. Cosmology thus differs from other sciences in that its subject matter is unique – there is only one universe. There is no greater challenge to science than to explain how the universe came into existence and how it achieved its present form and structure.

Modern cosmology began in the 1920s, with the discovery by Edwin Hubble that the universe is expanding. Most cosmologists now agree that this expansion is a remnant of a gigantic explosion, or big bang, which occurred about fifteen billion years ago and marked the origin of the universe in its entirety. Evidence for the big bang also comes from the existence of a uniform background of heat radiation which bathes the universe at a temperature of about 3 K. Discovered accidentally in 1965, this background heat is the last fading glow of the primeval inferno which marked the violent birth of the cosmos. Much of the effort of cosmological research in the last two decades has been directed towards securing an understanding of the stages of the universe immediately following

1. *This particle accelerator machine at CERN is capable of momentarily simulating the conditions that prevailed in the primeval universe less than one microsecond after the beginning.*

the big bang, and in trying to relate features observed in the universe today to physical processes that took place during that primeval phase.

How do we know that the universe is expanding? The most direct evidence comes from examining the quality of the light received from distant galaxies. It was found by Hubble and others that this light is systematically shifted towards the red end of the optical spectrum. This means that the light waves are stretched somewhat relative to similar sources of light on Earth. Such a 'red-shift' is familiar in the laboratory as a sign that the source of light is moving rapidly away from the observer, and this is how Hubble interpreted it. He concluded that the galaxies are rushing apart.

Galaxies can be thought of as the fundamental building blocks of cosmology. It is their recessional motion that defines the expanding universe. Within a galaxy there is no expansion. Our own Galaxy, the Milky Way, consists of about one hundred billion stars arranged in a flat disc, slowly orbiting the galactic core. The Milky Way is typical of a class of galaxies known as spirals, because they display spiral arms. Other forms are also known. There is a tendency for galaxies to aggregate together in clusters, attracted by their mutual gravitation. This tendency opposes the general expansion, so it is really more accurate to envisage the clusters rather than the individual galaxies as the basic cosmological units.

Hubble spotted that the fainter galaxies had the larger red-shifts. Because a galaxy appears fainter the farther away it is, he interpreted this to mean that the more distant galaxies are receding faster. Figure 3 shows a graph of optical magnitude plotted against red-shift for a sample of galaxies. These sorts of graphs are known, appropriately enough, as Hubble diagrams. The red-shift is here defined as the ratio of the received wavelength to the wavelength from a stationary standard source in the laboratory. Assuming that brightness is on average an accurate measure of distance, and red-shift really does represent recessional velocity, the graph can be regarded as a plot of distance against velocity, and

2. *Arno Penzias and Robert Wilson shown with the equipment that accidentally detected the heat radiation from the big bang.*

the units used have been adapted to show this. The diagram therefore indicates how fast a galaxy at a given distance is moving away from us.

The most striking feature of the Hubble diagram is that, allowing for the expected scatter in the data, the points lie along a straight line. This implies that a galaxy twice as far away recedes at twice the speed, a relationship known as the Hubble law. The pattern of expansion is therefore very regular. The slope of the line, which is called Hubble's constant, is a key cosmological parameter. Most astronomers accept a value for Hubble's constant of about 50 km/s/Mpc. This means that a galaxy one megaparsec (1 Mpc = 3.26 million light-years) away recedes at 50 km/s. A galaxy 10 Mpc away recedes at 500 km/s, and so on.

This simple linear relationship between distance and velocity has a deep implication for the nature of the expanding universe. It means that the universe is expanding at the same rate everywhere: viewed from any other galaxy, the pattern of motion would be more or less the same. It is wrong to imagine, as many people do, that we are somehow at the centre of the expansion. Although most other galaxies are certainly moving away

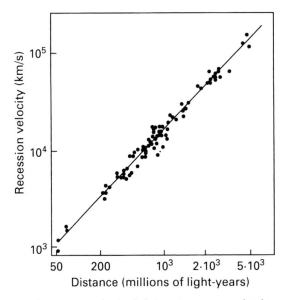

3. *Hubble diagram showing a graph of red-shift against apparent brightness for a selection of galaxies. The significant feature is that, in spite of the scatter, the points cluster along a straight line.*

from us, they are also on average moving away from each other, and because of the Hubble law the neighbours in the vicinity of any other galaxy recede from it in much the same way as our neighbouring galaxies recede from the Milky Way. No galaxy is in the privileged position of being at the centre of the expansion.

Some people have difficulty with this idea, and a helpful analogy is that of a rubber sheet covered in dots to represent the galaxies (see Figure 4). Imagine that the sheet is being stretched by pulling evenly around the edge. The effect is to cause every dot to move away from every other, exactly as in the expanding universe. Moreover, the system

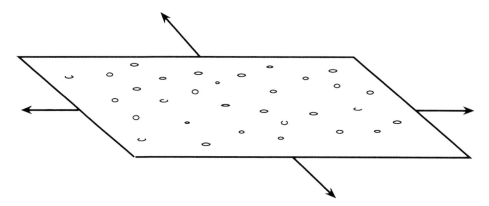

4. *Expanding space represented by analogy with a rubber sheet being stretched evenly in all directions. The dots represent galaxies.*

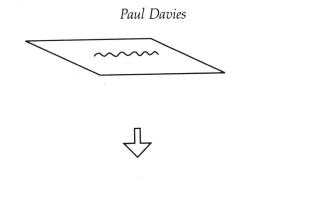

5. *The cosmological red-shift can be envisaged as produced by the expansion of space. Using the rubber sheet analogy, the length of a light wave is shown being stretched as the sheet expands.*

obeys the Hubble law: dots separated by twice the distance recede from each other at twice the speed.

Now it could be objected that the dots on the sheet *are* moving away from a common centre, namely the middle of the sheet. If, however, the sheet were so large that you could not see the edges, there would be no way of knowing which dots were near the middle, and which were not, merely by inspecting their relative motions. If the sheet were *infinite* in extent, then there would truly be no meaning to the concept of centre or edge. In the real universe there is not the slightest hint that the assemblage of galaxies has an edge anywhere, so there is no reason to talk about a centre to the universe, or a region away from which the galaxies are rushing.

Nevertheless, one is still tempted to ask whether there *really is* an edge out there somewhere, beyond the range of our current telescopes. After all, we cannot be certain that there are galaxies populating the universe all the way to infinity. Curiously, it is possible for the universe to be finite in extent, and yet still have no centre or edge. Leaving that possibility aside for now, there is a sense in which the speculation about a very distant edge to the universe is pointless, if not meaningless. The Hubble diagram shows that the recessional speed of the galaxies grows with distance, and there comes a point at which this speed is so great it reaches the speed of light. Observationally, what we see is a red-shift that grows progressively greater with distance, until the light is so red-shifted it lies outside the visible spectrum altogether.

Clearly we could not observe galaxies that recede faster than light, for their radiation could never reach us, either as visible light or in any other form. So we cannot see beyond a certain distance, however powerful our telescopes. The limit in space beyond which we cannot see, even in principle, is called our horizon. As with the terrestrial horizon, its existence does not mean that nothing lies beyond, only that whatever does, we cannot see it. In the conventional big bang model our horizon is located about fifteen billion light-years away, and our instruments can directly probe to within a million or so light-years of its location. No edge to the universe has been discerned. Any such edge that might exist beyond the horizon is in principle unobservable from Earth (at least at this epoch), so we might as well forget it. It is irrelevant to the *observable* universe.

There is another, rather fundamental, reason to believe that the universe is expanding. On a cosmological scale the dominant physical force is gravitation. Being attractive, gravitation tries to pull the cosmological material together. The universe could not remain static in the presence of such a universally attractive force; it would inevitably collapse. Only by flying away from each other can the galaxies avoid falling together.

It is obvious that if the galaxies are moving apart, they must have been closer together in the past. Extrapolating this trend, it seems that there must have been a time when all the matter in the universe was compressed together. Knowing the rate of expansion, we can estimate when this dense phase was. According to the Hubble law the universe doubles in size every several billion years, so 'running the movie backwards' we might expect that several billion years ago the universe was in such a dense state.

There is, however, a subtlety. The gravity of the universe acts on the retreating galaxies and restrains their dispersal. Thus the rate of expansion of the universe is continually slowing. Conversely, the universe expanded at a faster rate in the past. We must take this deceleration into account when estimating how long ago it was that the universe was highly compressed. Note that as the expansion rate slows, so galaxies that start out beyond our horizon eventually come within it. What happens is that the horizon itself expands *faster* than the universe, so that as time goes on it encompasses more and more galaxies — even though the galaxies continue to retreat from us.

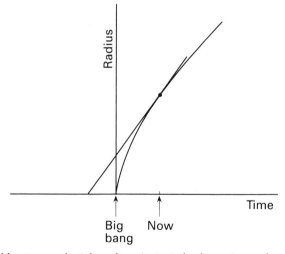

6. *The curved line is a graph of the radius of a typical volume of expanding space plotted as a function of time. The tangent to the curve at the present epoch represents the graph that would be obtained if the gravitational deceleration of the universe were ignored. The point of intersection of the curve with the horizontal axis corresponds to the big bang. The volume of space apparently goes to zero there. This is called a space-time singularity.*

Figure 6 shows how the radius of a typical volume of space varies with time. The curve should be compared with the straight line, which is what would be obtained if the deceleration were ignored and the present rate of expansion were assumed for all cosmic epochs. The slope of this line is determined by the Hubble constant. Two things are clear from Figure 6. The first is that a constant rate of expansion overestimates the duration that has elapsed since the compressed phase by a factor of 3/2 or so. Secondly, the curve suggests that the rate of expansion must once have been very rapid indeed compared to its value today. We are thus led to a picture in which the universe that we know emerged several billion years ago from a dense phase with an explosively rapid expansion. This is why the big bang is so-called.

A deep puzzle concerns the form of the explosion. Observations of the background heat radiation show that on a large scale the universe is extraordinarily uniform in the way that matter and energy are distributed throughout space and in the rate at which the expansion occurs. 'Large scale' here means on sizes greater than a super-cluster of galaxies (about one hundred million light years). This uniformity implies that the universe would look much the same from any other galaxy as it does from our own. There is nothing special or privileged about our location in the universe. Moreover, the uniformity is maintained with time, so our Galaxy shares a common cosmological experience with other galaxies as the epochs go by. Evidently the big bang was a highly orchestrated outburst in which different regions of the universe began expanding at precisely the same moment with the same degree of vigour everywhere and in all directions.

This remarkable synchronisation is all the more baffling when account is taken of the theory of relativity, which forbids any physical influence to exceed the speed of light. Our telescopes can today probe different regions of the universe on opposite sides of the sky which apparently lie beyond each others' light 'horizons'; that is, although light has been able to reach us from these regions, it has not yet had time to cross from one to the other. This means that such regions cannot have ever been in causal contact, yet they have somehow contrived to expand at the same rate and produce matter at the same average density.

Yet cosmic uniformity is not quite exact. On a relatively 'small' scale, matter is clumped into galaxies and galactic clusters, the origin of which is still not fully understood. Astronomers have calculated that for this local structure to exist today, some sort of irregularities must have been present in the universe at the outset to act as 'seeds' or condensation centres around which galaxies could grow. It would therefore appear that the universe began in a rather peculiar state of order, in which large-scale regularity and local irregularity were present together.

Had the big bang been less violent, the cumulative gravity of all the cosmic material would have caused the entire universe to fall back on itself after a brief duration and collapse in a 'big crunch'. Alternatively, had the bang been more violent, the cosmic material would have dispersed so quickly that galaxies would never have formed and the average density of matter in the universe would now be minute. For some mysterious reason the vigour of the explosion was matched to the gravitating power of the universe so delicately that the galaxies have just enough speed to escape each other's gravity, yet not so much as to produce rapid dispersal. The universe seems to lie precisely on the borderline between the two alternatives of too much gravity or too much violence.

If we go back to 10^{-43}s, which is the earliest moment at which it makes any sense to talk about cosmic dynamics, we find that the matching of explosive vigour to gravitating power was accurate to no less than one part in 10^{60}. This astonishing fidelity is perplexing. Why should the universe be so propitiously arranged to such phenomenal accuracy?

Radio telescopes are able to prove much earlier epochs than optical telescopes. Indeed, it is possible to detect radio waves that come to us from an epoch less than one million years after the big bang, with a red-shift of about 1000. Remember, one million years is less than one-thousandth of the age of the universe. At that epoch galaxies had not yet formed and the universe was more or less evenly filled with hot hydrogen and helium gas. Observations thus show a uniform background of heat radiation which bathes all space at a temperature of about 3 K ($\approx -270°$C).

Like most physical systems the universe cools as it expands. The background heat radiation was therefore at a higher temperature in the past. At one million years the temperature was around 4000°C. Although one cannot directly observe the radiation emanating from epochs before that time, due to the obscuring effects of the ionised gas, it is possible to extrapolate mathematically to deduce that, say, one second after the big bang the temperature was ten billion degrees. This is so hot that none of the familiar structures we now find in the universe, including atomic nuclei, could have existed. It leads to the conclusion that the cosmic material that emerged from the big bang started out in the form of a soup of unbound subatomic particles, mainly protons, neutrons and electrons. Thus the primeval universe was not only very dense and rapidly expanding, it was also exceedingly hot.

In the 1960s cosmologists realised that temperatures were so high during the first few minutes after the big bang that nuclear reactions would have occurred among the particles in this soup. Detailed calculations were performed, and it was found that all the primeval neutrons and a large fraction of the protons would have combined to form nuclei of the element helium; the remaining protons would have gone on to form hydrogen. The prediction was made that these chemical elements ought to have an abundance by weight of about four parts hydrogen to one of helium. Observations confirm that the universe is indeed composed of about 75% hydrogen, 25% helium, and only minute traces of heavier elements. This was a great triumph for the big bang theory and confirmed that the element helium is a relic of the early universe, originating from the first few minutes after the explosive beginning.

But what about the protons, neutrons and other particles present in the primeval soup? What is their origin? Ever since Albert Einstein wrote down his equation $E = mc^2$, it has been known that matter is a form of encapsulated energy. It follows that energy can be used to create matter. It is therefore tempting to suppose that the primeval soup, and therefore all the cosmic material we see today, was produced from the energy of the big bang.

There is, however, a snag. Whenever matter is produced in the laboratory it is always accompanied by an equivalent mass of antimatter, which is a sort of mirror image of matter. Because matter and antimatter annihilate instantly on contact to produce intense gamma radiation, astronomers believe that there can be only minute traces of antimatter in the universe. The problem is then, why is there so little cosmic antimatter?

One theory is that protons, which had hitherto been assumed to be absolutely stable, can decay with a very low probability, corresponding to an average lifetime of at least 10^{32} years. This process, though exceedingly weak, breaks nature's symmetry between matter and antimatter. To see this, imagine the fate of a hydrogen atom in which the proton decays to a particle called a pion, plus a positron (the antimatter partner to the electron). The pion soon decays into photons, while the positron and electron could annihilate into more photons. Thus the material of the hydrogen atom has turned into pure radiation energy. The reverse of this process would be the creation of matter without antimatter from heat radiation.

Calculations suggest that in the big bang something like this actually occurred, but it was greatly speeded up by the extreme conditions. At about 10^{-32}s, known as the unification epoch, huge quantities of energy were abruptly released to become heat, matter and all the other forms of cosmic energy that power the universe today. Because the matter–antimatter symmetry breaking is so weak, however, only a tiny fraction of the cosmic material would have been created by this mechanism. Mostly, matched pairs of particles and antiparticles would have been produced. As the universe cooled, these pairs annihilated into photons, leaving only a minute residue – about one part in a billion – of the excess particles. It is from this residue that all the galaxies are made. The annihilation photons eventually cooled to form the background heat radiation that was discovered in 1965. In a sense, then, the matter that constitutes our bodies is a relic from the unification epoch, 10^{-32}s after the creation.

Just how far back can a scientific account of the universe take us? Look again at Figure 6. If it is taken literally, and one follows the curve back to the beginning, then the volume of a typical region of space goes to zero. This implies that the universe was infinitely compressed, with all the cosmic material squashed into a single point. Cosmologists use the term 'singularity' to denote this limiting state. According to the general theory of relativity, a singularity represents a boundary to space and time. Here, all physical theories break down. It is not possible to continue space and time back through the singularity. In other words, one might say that space and time *came into existence* at the singularity. For this reason, the big bang is taken to represent the origin of the entire physical universe, and not merely the origin of matter.

Most people think of space and time as simply *there*. The idea that they may not always have existed is hard to grasp. Perhaps the greatest lesson of twentieth-century physics, however, is that space and time are not merely an eternally existing arena in which the drama of the universe is acted out: they too are part of the cast. And the coming into being of the universe in the big bang involved more than the abrupt appearance of matter and energy. It was the coming into being of space and time as well. The concept of an origin to time is not new. In the fifth century St Augustine of Hippo proclaimed that the world was made *with* time and not *in* time. If this is correct then the oft-repeated question 'What happened before the big bang?' becomes meaningless. There was no 'before'.

A related question is: Where did the big bang occur? The answer: everywhere! Remember that there is no centre or edge to the universe. The explosion did not occur

7. *This proton–antiproton collider at CERN is a factory for producing antimatter and studying its interaction with ordinary matter.*

in space, at some particular location. The big bang was the explosive appearance *of* space. To make this point clearer, let us return to the rubber sheet analogy. This time, however, the rubber sheet is curved as, for example, in a child's balloon. The balloon is covered more or less uniformly in dots to represent galaxies (see Figure 8). Imagine the radius of the balloon being progressivly shrunk; this corresponds to going backwards in time towards the big bang. Remember that the fabric of the balloon represents space itself. As the balloon gets smaller, so there is less and less space. In the limit that the balloon shrinks to zero radius, the volume of space dwindles to nothing and the 'universe', space and all, simply disappears at a point. According to this analogy, then, the big bang was the abrupt creation of the universe from literally nothing: no space, no time, no matter.

We are led to this extraordinary picture of a physical universe that simply pops into existence from nothing by taking Figure 6 seriously. The curve is based on an idealised model for the universe in which matter is spread with precise uniformity and gravity is always attractive. If the universe is exactly uniform and it is evenly shrunk to the limit, it is obvious that all the matter must be squashed to the same point. However, the universe is not precisely uniform: matter is aggregated into galaxies, for example. Moreover, the rate of expansion cannot be *exactly* the same everywhere and in all directions. It seems at first that these imperfections must invalidate the conclusion that there was a singularity marking a past limit to the universe, for if one were to follow back in time the careers of different portions of the universe then they would surely not come together in *exactly* the same place at the same time. Remarkably, though, it can be shown that even in an irregular universe a singularity (of some more complicated sort) is inevitable, so long as gravity is always attractive.

Some cosmologists have conjectured that perhaps, under the extreme conditions of the big bang, something like antigravity is possible. This could remove the singularity. A possible scenario would then be something like this. Before the big bang the universe was contracting, falling together under its own gravity. At some stage when the density was

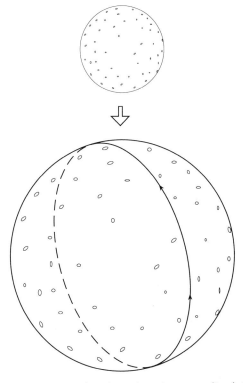

8. *A spatially closed universe depicted as the surface of an expanding balloon. Also shown is the path of a hypothetical explorer who circumnavigates the universe by travelling always in the same direction.*

very high, gravity turned into antigravity, causing the universe to 'bounce', thereby initiating the present phase of expansion.

But now a new problem presents itself. If the universe did not come into existence a finite time ago at a singularity, then it must always have existed. This means that physical processes must have been going on for all eternity. But essentially all the physical processes that we observe in the universe are finite and non-renewable. Stars, for example, do not shine for ever. Eventually they burn up all their fuel, perhaps collapsing to form a black hole. The supply of material for new stars is limited, so these irreversible processes cannot have been going on forever.

It might be countered that the high-density 'bounce' phase would pulverise and reprocess matter, destroying all trace of earlier systems and structures, and thereby renewing the universe. There is, however, a fundamental law of physics, called the second law of thermodynamics, which places strong constraints on what can be achieved by a cyclic physical process. In particular, it forbids any process from returning the universe as a whole to an earlier physical state. For these reasons most cosmologists prefer to believe that the universe has a finite age, and that the big bang really did mark a creation from nothing.

Even if the universe was created from nothing, it is still tempting to ask what caused such an event. We are used to the fact that events are caused by earlier physical processes, but in the case of the origin of the universe there cannot have been any prior physical processes because the big bang was the beginning of time. Nevertheless there is still a strong urge to ask *why* the universe suddenly appeared from nothing.

For a long while cosmologists dismissed the question as lying outside the scope of science. The big bang, they replied, 'simply happened'. Many people find this response unsatisfactory. Some have tried to evade the problem by insisting on a universe of infinite age. However, quite apart from the problems with an eternal universe already discussed, this strategy is really no help. One can still ask why an eternal universe exists, rather than no universe at all.

In the past ten years a number of cosmologists have come to believe that they have solved the problem of how the universe came to exist from nothing. The key is to appeal to quantum physics. Quantum mechanics is normally restricted in its application to atomic-scale systems because quantum effects in macroscopic objects are negligible. But in the extreme conditions of the big bang they cannot be neglected. A central feature of quantum mechanics is that all physical quantities are subject to unpredictable fluctuations. In quantum cosmology these fluctuations involve space and time. It has long been conjectured that, at an ultra-microscopic level, space and time are subject to chaotic unpredictable changes. According to these speculative ideas, space can stretch and curve so violently that it forms a complex architecture of 'wormholes', tunnels and bridges. This shifting complexity is sometimes referred to as space-time foam. Amid this 'foam', minute self-contained universes are continually being created and destroyed by quantum fluctuations.

According to the latest thinking, the universe began with such a minute proto-universe. It measured a mere 10^{-33} cm in size. This is known as the Planck length, after Max Planck, the originator of quantum physics. The associated time scale which characterises change at this length is 10^{-43}s, the Planck time. It is not necessary to account for how the proto-universe fluctuated into existence: it is the very essence of quantum mechanics that such fluctuations are unpredictable even in principle. This inherent indeterminism means that the unannounced appearance of a mini-universe is an event entirely consistent with the laws of quantum physics.

The problem, then, is not to explain how the proto-universe came to be, but how it developed from a minute 'whimper' into a big bang. Somehow the nascent universe boosted its size from almost nothing to literally cosmic proportions. It had to do this very rapidly, before it could collapse back out of existence once more (as do the vast majority of the fluctuations, such as those which are occurring within the space-time foam at this very moment). One idea to explain this promotion from whimper to bang is called the inflationary scenario. The essential idea is that, for a fleeting moment, the proto-universe was dominated by antigravity, which caused it to expand at an ever faster rate.

Curiously, just such a repulsive or antigravitational force was invented by Einstein as early as 1917. Although Einstein himself later discarded the cosmic repulsive force, and no astronomer has yet detected any sign of it existing today, it fits in very well with the general theory of relativity. In the last few years, many physicists have come to believe that some sort of cosmic repulsion was actually unavoidable in the early universe, as a by-product of the other forces of nature.

The strength of the repulsive force could have been truly enormous — at least one hundred powers of ten larger than the observed upper limit on its strength today. This discrepancy would be a major headache if it were not for the fact that the repulsion is temperature-dependent. It is possible for the repulsive force to be virtually zero at our present, low-temperature epoch, and yet to have completely dominated the behaviour of the universe at around 10^{-35}s, when the temperature was about 10^{28}K.

It was pointed out in 1981 by Alan Guth of Massachusetts Institute of Technology that the repulsion force could propel the universe into a phase of runaway expansion and solve at a stroke many of the problems that plague modern cosmology. Seized in the grip of the powerful repulsion the universe would have inflated exponentially fast, doubling its size every 10^{-35}s, or so. At this dizzy rate, it would have taken but the twinkling of an eye for the universe to swell to a gigantic size.

The significance of this astronomical distension is not only that it could convert a minute blob of spontaneously generated space into a veritable universe, but that any initial irregularities present, such as turbulence or an uneven distribution of energy, would be hugely diluted and smoothed away by the colossal stretching involved. We might therefore expect the universe to emerge from its inflationary phase with a highly uniform distribution of matter and motion. That is precisely what is observed.

Guth's original inflationary scenario has undergone several refinements, and now offers the promise of explaining the origin of galaxies too. The inflation would not have been precisely uniform, because it is essentially a quantum effect and all processes of quantum physics involve random fluctuations. These would have caused some regions to inflate slightly more than others, producing a texture of local small-scale irregularities which would then encourage the growth of galaxies.

Moreover, the inflationary mechanism avoids the light horizon problem because the inflationary phase displaces the location of the horizon far beyond the fifteen million light-years that we assumed on the basis of the conventional theory. As a result, the size of the regions today which were in causal contact prior to inflation is much greater than the currently observable universe.

Whatever the vigour of the initial bang, its effect is utterly swamped by the rising tide of inflation. At the end of the inflationary phase the universe has entirely forgotten its original activity, and the behaviour stamped on the subsequent epochs carries only the imprint of the inflation. It happens that exponential inflation has a curious property. It delivers to the emergent universe precisely the rate of expansion that corresponds to the exact matching of gravitating power to explosive vigour discussed earlier. Another mystery is solved.

This catalogue of successes has endeared the inflationary scenario to many cosmologists. It is not, however, without its problems. Foremost among these is the question of how the runaway inflationary phase comes to an end and returns the universe to the more conventional activity of gradually slowing expansion. For inflation to work properly, it must be sustained for long enough for the universe to double in size at least eighty five times. During this time the universe swells by many powers of ten and its temperature falls by the same factor, dropping almost to absolute zero. Thus, the universe more or less instantaneously cools from an enormous temperature to nothing. The way then lies open for the universe to drop into its low-temperature phase, in which the cosmic repulsion force disappears.

This change of phase, akin to the transition from water to ice, will obviously bring about the end of the inflation by removing its driving force. To avoid this happening too quickly, Guth and his colleagues proposed that the cosmic material underwent a period of supercooling. When pure water is carefully cooled, it can remain liquid somewhat below freezing point. A slight disturbance will cause the supercooled liquid to abruptly solidify into ice. By analogy, the high-temperature, grand-unified phase of the universe could have persisted for a while after the temperature had dropped away to nothing as a result of the inflation, enabling the necessary amount of swelling to take place before the 'freeze-out' occurred.

The chief difficulty with this idea has to do with the details of the freeze-out. Studies by Sidney Coleman of Harvard University suggested that such a phase transition would occur by means of nucleation. Roughly speaking, small 'bubbles' of the new phase appear at random, and start to grow with the speed of light, eventually intersecting each other and filling all of space. Inside the bubbles inflation comes to a shuddering halt. The energy of the runaway expansion is transferred instead to the bubble walls. When these highly energetic walls start to collide, they would dissipate their energy rapidly in the form of heat, giving back the vast reserves of thermal energy that were sapped from the cosmos when the inflationary phase began. Thus the universe is returned abruptly and explosively to a high temperature once more, but this time without the repulsive force. From this condition it is free to continue along the conventional path of decelerating expansion from a hot state, with the uniformity, horizon and expansion rate problems attended to.

The snag in all this concerns the collisions between the bubble walls. Wall encounters would be random and chaotic, and would seem to introduce a large amount of turbulence and inhomogeneity, the very things the inflationary scenario was designed to ameliorate. There is as yet no unanimous agreement about the best way to avoid this difficulty, which has become known as the 'graceful exit' problem, but a number of possibilities have been suggested.

A promising proposal is that, rather than appeal to supercooling as a cause of the cosmic hang-up during the inflationary period, perhaps the phase transition could simply have been a rather sluggish process. The essential idea can be illustrated by analogy. Imagine a ball resting on top of a hill. The system is unstable because a slight disturbance will cause the ball to roll off the hill down into the valley where it can come to rest in stable equilibrium. Here the top of the hill represents the inflationary phase when antigravity dominates, whilst the valley represents the present phase in which gravity is dominant. Obviously freeze-out will take a finite amount of time, corresponding to the time taken for the ball to roll down the hill; it is not an instantaneous process. If the slope of the hill near the top were very shallow, the ball would roll only slowly at the outset, and very little change would occur at first. There is some hope that the quantum processes that drive the cosmic phase transition will behave in just this way, delaying the freeze-out long enough for the inflation to work its magic, but avoiding the problems of bubble formation that supercooling would cause.

Assuming all this can be made to work, there still remains a rather deep philosophical problem. The difference in the strength of the cosmic repulsion between the high- and low-temperature phases is, as we have seen, at least one hundred powers of ten larger than the maximum current value permitted by observation. Why the low-temperature value turns out to be so close to zero is completely unexplained by the inflationary theory. It is a circumstance made all the more mysterious by the fact that there are probably several contributions to the total strength of the repulsion. Evidently some pretty exquisite cancellations are occurring for as yet obscure reasons.

Cosmologists call this 'the cosmological constant problem', and it seems likely to attract the attention of theorists in years to come. Some partly formulated ideas concerning 'wormholes' and space-time 'foam' suggest that quantum gravitational processes might force the above-mentioned cancellation exactly. No doubt other proposals will be made as new ideas are developed for the unification of gravity with nature's other forces.

On the observational front, we have yet to discover any relic that gives a clear indication of the state of the universe during the first one second. Observers are hopeful that experiments may soon be constructed to detect the 'dark matter' that may make up 90% or more of the mass of the universe. At present the nature of this invisible stuff is a mystery, but it seems likely that at least part of it is left over from the very early universe.

As a result of recent observations of the large scale structure of the universe, some cosmologists are beginning to question the simple big bang picture that I have outlined in this chapter. The problem concerns the fact that although on very large length scales the distribution of matter in the universe is more or less uniform, astronomers are discovering more and more structures on medium length scales. In some cases thousands of galaxies cluster together to form huge sheets and filaments. One of these features, known as the Great Wall, stretches an appreciable distance across the visible universe. It is becoming increasingly clear that the cosmos has a sort of frothy appearance, with vast aggregations of matter surrounding apparently featureless voids.

There is difficulty reconciling this clumpiness in the distribution of matter with the very high degree of smoothness of the cosmic microwave background radiation. The Cosmic Background Explorer (COBE) satellite is currently probing this radiation, and the results show no sign of anisotropy to at least one part in ten thousand. If, as the standard scenario requires, the radiation is a relic from the early universe, it should carry the imprint of any irregularities that were present in the distribution of matter at about one million years after the big bang. As no such irregularities are yet apparent, it seems as if the universe in its early stages was remarkably uniform. The problem is to explain how the present clumpy state of the universe has emerged from such smooth beginnings given the relatively limited time that has elapsed since the big bang. Attempts to model the growth of clumpiness on the assumption that there are large quantities of dark matter have difficulty in explaining the precise form of the observed clumping, at least for simple types of dark matter. It may be that a more complicated mixture of dark matter can provide a satisfactory explanation, or that the standard big bang scenario is flawed in some more fundamental way.

Meanwhile, refinements are sure to be made to the inflationary scenario, which is still in its infancy. Many of the quantum details are hard to analyse and are highly model-dependent. It is far too soon to pronounce the theory a complete success. Yet it contains several features that so neatly account for otherwise mysterious cosmological facts that it is hard to resist the impression that some sort of inflationary activity must have occurred during that first brief flash of existence.

If inflation can be made a success, it provides a mechanism for converting a tiny quantum universe into a full-blown cosmos, and enables us to contemplate scientifically the creation *ex nihilo* of theology. A tiny bubble of space pops spontaneously and ghostlike into fleeting existence as a result of quantum fluctuations, whereupon inflation seizes it and it swells to macroscopic dimensions. 'Roll-over' then occurs and the expansion drops in rate amid a burst of heat. The heat energy and gravitational energy of expanding space then produce matter, and the whole assemblage gradually cools and decelerates to the condition we observe today. Hey presto, a universe!

The great Roman philosopher Lucretius (99–55BC) once proclaimed that 'Nothing can be created out of nothing'. Cosmologists are slowly beginning to realise that perhaps everything can be created out of nothing.

15

Origin of the universe

MARTIN J. REES*

Everyone is familiar with the idea that evolutionary processes – geological and biological – have led to the present state of the Earth and the creatures on it. Astrophysicists and cosmologists have a broader perspective: they aim to set the Earth itself in a cosmic evolutionary context, and trace the origins of its constituent material back to the formation of our Milky Way Galaxy – maybe even right back to the first seconds of the so-called 'big bang' that initiated our expanding universe. They cannot yet offer more than the rudiments of this overall cosmogonic scheme; but what should really surprise us is that there has been any progress at all. 'The most incomprehensible thing about the universe is that it is comprehensible' is one of Einstein's best-known sayings. It expresses his wonder that the physical laws, which our brains are somehow attuned to make sense of, apparently apply not just in the laboratory but in the remotest parts of the universe. This unity and inter-relatedness of the physical world must impress all who ponder it. Later on, we shall venture towards some speculative fringes of the subject, but let us start with something quite well understood, the life cycle of a star such as our own Sun.

The Sun and stars

The Sun started life by condensing gravitationally from an interstellar cloud. It continued to contract until its centre became hot enough to ignite nuclear reactions. Gradual

* This contribution is an adapted version of my article 'Origin of the Universe' which appeared as Chapter 1 of *Origins* (ed. A. C. Fabian), Cambridge University Press, 1988.

conversion of hydrogen into helium then releases enough energy to keep the Sun burning steadily, as a gravitationally confined fusion reactor, for about ten thousand million (ten billion) years. It has been shining for four-and-a-half billion years, and about five billion years from now the hydrogen in its core will run out. It will then swell up to become a red giant, engulfing the inner planets, before settling down to a quiet demise as a white dwarf.

The study of stellar evolution made little progress until the 1930s. Before that time, the physics of nuclear reactions was not understood. Lord Kelvin argued that gravitation must be the prime energy source, and that the Sun must inexorably contract as it loses heat. He calculated that unless sources now unknown to us are 'prepared in the great storehouse of creation', the Sun could last only twenty million years. Only much later did laboratory physics reveal the nuclear fuel that Kelvin could not conceive of.

Stars are so long lived compared with astronomers that we only have, in effect, a single we *can* check our theories, just as we could infer the life cycle of a tree by one day's observation of a forest. Of special interest are places like the Orion Nebula (Figure 1), where even now stars, perhaps with new solar systems, are condensing from glowing gas

1. The Orion Nebula – a region where stars are now forming

2. A globular star cluster — a self-gravitating system of several hundred thousand coeval stars. This view of M22, in the constellation of Sagittarius, was photographed by Akira Fujii.

clouds; and star clusters (Figure 2), containing stars of different sizes which are thought to have formed at the same time.

Not everything in the cosmos happens slowly. Stars heavier than the Sun evolve faster, and some expire violently as supernovae. The best-known instance is the Crab Nebula, the expanding debris of a stellar explosion seen and recorded by oriental astronomers in AD 1054. In July of that year, Yang Wei Te, the Chinese 'chief calendrical computer' (the counterpart of our Astronomer Royal, presumably), reported to the Emperor that a 'guest star' had appeared. This star faded after a few months, leaving a remnant behind. Supernova explosions signify the violent end-point of stellar evolution, when a star too massive to become a white dwarf exhausts its available nuclear energy. The star then faces an energy crisis. Its core catastrophically implodes, releasing so much gravitational energy that the outer layers are blown off. The centre of the star collapses to form a spinning neutron star (pulsar) only about 10 km across.

Supernovae may seem remote and irrelevant to our own origins. But on the contrary, only by studying the births of stars, and the explosive way in which they die, can we tackle such an everyday question as where the atoms we are made of came from. The respective abundances of the elements of the periodic table can be measured in the solar system, and inferred spectroscopically in stars and nebulae. The proportions in which the

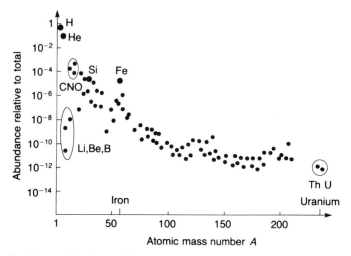

3. *The abundances of the chemical elements in the solar system are here plotted (on a logarithmic scale) as a function of atomic number. Note that only 2% of all the matter is in elements heavier than hydrogen and helium. No objects have been found to contain less than 23–25% (by mass) of helium. It is believed that the material emerging from the big bang was essentially just hydrogen and helium, and that heavier elements were synthesised in stars over the lifetime of the galaxy.*

elements occur display regularities from place to place which certainly demand some explanation (Figure 3).

Complex chemical elements are an inevitable by-product of the nuclear reactions that provide the power in ordinary stars. A massive star develops a kind of onion-skin structure, where the inner hotter shells are 'cooked' further up the periodic table (Figure 4). The final explosion ejects most of this processed material. The relative abundances of the heavy elements are fairly standardised throughout the Galaxy; in some of the oldest stars, however, these elements are all depleted relative to hydrogen and helium compared to abundance measurements in the solar system. All the carbon, nitrogen, oxygen and iron on the Earth could have been manufactured in stars that exhausted their fuel supply and exploded before the Sun formed. The solar system would then have condensed from gas contaminated by the debris ejected from earlier generations of stars. These processes of cosmic nucleogenesis can account for the observed proportions of different elements – why oxygen is common but gold and uranium are rare – and how they came to be in our solar system.

Each atom on Earth can be traced back to stars that died before the solar system formed. A carbon atom, forged in the core of a massive star and ejected when this exploded as a supernova, may spend hundreds of millions of years wandering in interstellar space. It may then find itself in a dense cloud which contracts into a new generation of stars. It could then be once again in a stellar interior, where it is transmuted into a still heavier element. Alternatively, it may find itself out on the boundary of a new solar system in a planet, and maybe eventually in a human cell. We are literally made of the ashes of long-dead stars.

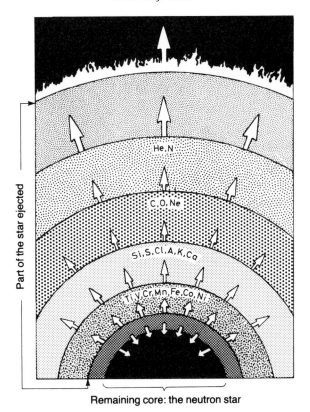

4. The structure of a massive star before the final supernova outburst. The hotter inner shells have been processed further up the periodic table; this releases progressively more energy until the material is converted into iron (A = 56), the most tightly bound nucleus. Endothermic nuclear reactions occurring behind the shock wave that blows off the star's outer layers (explosive nucleosynthesis) can synthesise small quantities of elements beyond the 'iron peak' (see Figure 3).

This concept of *stellar nucleosynthesis*, due primarily to Fred Hoyle, Willy Fowler and Alisdair Cameron, is one of the triumphs of astrophysics in the last forty years. It sets our solar system in a kind of ecological scheme involving the entire Milky Way Galaxy. The particular mix of elements that we find around us is not *ad hoc*, but the outcome of transmutation and recycling processes, whose starting point is a young galaxy containing only the lightest elements. One thing this scheme could not explain, however, was helium – why this makes up so much of the mass of all stars, young or old. We shall come back to that later.

Galaxies and their active nuclei

Let us now extend our horizons to the extragalactic realm. It has been clear since the 1920s that our Milky Way, with its 10^{11} stars and scale of about a hundred thousand

5. A disc galaxy, viewed at an oblique angle.

light-years, is just one galaxy, similar to millions of others visible with large telescopes.

Galaxies are held in equilibrium by a balance between gravity, which tends to draw the stars together, and the countervailing effect of the stellar motions, which if gravity did not act would cause the galaxy to fly apart. In some galaxies, our own among them, stars move in nearly circular orbits in giant discs (Figure 5). In others, the less photogenic

6. The constellation of Virgo contains a major cluster of galaxies including the two ellipticals M84 and M86 which are at centre and left of this picture.

ellipticals (Figure 6), the stars are swarming around in more random directions, each feeling the gravitational pull of all the others.

Our understanding of galactic morphology is tentative, maybe at the same level as the theory of individual stars was fifty years ago: there is much boisterous debate and several competing theories, but little consensus on the details. There is, however, a widely adopted scenario that accounts qualitatively for the two basic types – discs and ellipticals. Let us suppose that all galaxies started their lives as huge turbulent gas clouds contracting under their own gravitation, and gradually fragmenting into stars. The collapse of such a gas cloud is highly dissipative, in the sense that any two globules that collide will radiate their relative energy via shock waves, and will merge. The end result of the collapse of a rotating gas cloud will be a disc. This is the lowest energy state that such a cloud can reach if it loses energy but not its angular momentum. *Stars*, on the other hand, do not collide with each other, and cannot dissipate energy in the same fashion as gas clouds. This suggests that the *rate of conversion of gas into stars* is the crucial feature determining the type of galaxy that results. Ellipticals will be those in which the conversion is fast, so that most stars have already formed before the gas has had time to settle down in a disc. On the other hand, the disc galaxies will be those in which most star formation is delayed until the gas has already settled into a disc. The origin of these giant gas clouds is a cosmological question. But, given such clouds, the physics that determines galactic morphology is nothing more exotic or highbrow than Newtonian gravity and gas dynamics. This does not, of course, make the phenomena easy to quantify, any more than weather prediction is easy.

Some peculiar galaxies, though, are more than just a 'pile' of stars, and harbour intense superstellar activity in their centres. Nearby galaxies such as Centaurus A show this phenomenon in a mild way. But most extreme are the quasars, where a small region no bigger than the solar system outshines the entire surrounding galaxy, and the so-called

7. *The distribution over the sky of the brightest two million galaxies. Each dot represents a small patch of sky containing many images. Dots are black where there are no galaxies, white where there are more than twenty, grey for up to nineteen.*

radio galaxies, whose most conspicuous output is not visible light, but radio waves. In such galaxies, where the central power exceeds a million supernova explosions in unison, gas and stars have accumulated in the centre, until gravity overwhelms all other forces and a *black hole* forms. Here we need more highbrow physics, Einstein's general relativity (the theory that 'matter tells space how to curve, and space tells matter how to move'). The genesis of this theory was unusual: Einstein was not motivated by any observational enigma but, rather, by the quest for simplicity and unity; it was invented almost prematurely in 1915, when any prospects of observing strong field gravity seemed remote. But ever since the 1960s, when active galaxies were first fully investigated, relativists have been, in Thomas Gold's words 'not merely magnificent cultural ornaments, but actually relevant to astrophysics'.

The renaissance in general relativity in the 1960s stemmed not only from observational advances, but also from the deployment (by Roger Penrose, Stephen Hawking, and others) of novel mathematical techniques. Gravitational collapse, however asymmetrically it occurred, was found to lead to black holes whose properties could be exactly specified in terms of just two parameters: mass and spin. To quote Subrahmanyan Chandrasekhar: 'The black holes of nature are the most perfect macroscopic objects there are in the universe: the only elements in their construction are our concepts of space and time. And

222

since the general theory of relativity provides only a single family of solutions for their descriptions, they are the simplest objects as well.'

Black holes have now entered the general vocabulary, if not yet the common understanding. They are objects whose gravitational field is so strong – where space is so strongly curved – that not even light can escape. Black holes are the 'ghosts' of dead stars or galaxies – objects that have collapsed, cutting themselves off from the rest of the universe, but leaving a gravitational imprint frozen in the space they have left. To physicists, gravitational collapse is important because the central 'singularity' must be a region where the laws of classical gravitation are transcended, and one needs some unified physical theory to understand what is going on. Black holes bear on our general concepts of space and time because near them space behaves in peculiar ways that are highly non-intuitive to us. For instance, time would stand still for an observer who managed to hover or orbit just outside a hole's surface, and he could witness the entire future of the external universe in what to him seemed quite a short period. Even stranger and less predictable things might happen if one ventured inside.

Most theorists believe that the central prime mover in active galaxies involves a spinning black hole as massive as a hundred million Suns, fuelled by capturing gas, or even entire stars. This captured debris swirls downward into the hole, carrying magnetic fields with it and moving nearly at the speed of light. At least 10% of the rest mass energy of the infalling material can be radiated; further energy can be extracted from the hole's spin. Some of us are hopeful that these ideas can be put on a firm quantitative basis, just as our theories of stellar evolution have been. If so, this would offer real opportunities to probe the properties of strong gravitational fields where relativistic effects could be crucial.

Black holes were, in essence, conjectured more than 200 years ago. John Michell, an under-appreciated polymath of eighteenth-century science, published a paper in *Philosophical Transactions of the Royal Society* in 1784. He noted that the escape velocity from the Sun is about $\frac{1}{500}$ the speed of light, but would be 500 times larger for an object having the same density as the Sun but 500 times its radius. He noted that light would be made to return towards such an object 'by its own proper gravity'. Pierre Laplace made the same point a decade later, but if we accord Michell his deserved priority, 1984 was the black hole bicentennial. Of course, Michell's argument used the ballistic theory of light and Newtonian gravity, not the relativity that is really needed.

The expanding universe

The other arena where Einstein's theory is crucial is cosmology, the description of our universe as a single dynamical entity. Scientific cosmology is the study of a unique object and a unique event. No physicist would happily base a theory on a single unrepeatable experiment. No biologist would formulate general ideas on animal behaviour after observing just one rat, which might have psychoses peculiar to itself. But we plainly cannot check our cosmological ideas by applying them to other universes. Nor can we repeat or rerun the past, although the finite speed of light allows us to sample the past by looking at very remote objects.

Despite having these things stacked against it, scientific cosmology has proved imposs-

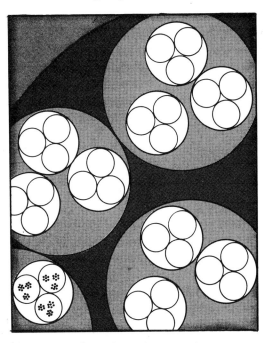

8. Charlier's universe: an infinite hierarchy of clusters. Such a universe would never look isotropic, however deep our surveys went.

ible, but only because the observed universe, in its large-scale structure, is simpler than we had any right to expect. It is of course sensible methodology to start by making simplifying assumptions about homogeneity, symmetry, etc., and cosmologists did this: indeed, back in the 1920s, Alexander Friedman, Georges Lemaître and others devised cosmological models based on Einstein's relativity. But what *is* surprising is that these models remain relevant, and the simplifying assumptions have been vindicated.

We are talking now about vast intergalactic scales of distances. With a large telescope, one can probe billions of light-years into space. To the cosmologist, even entire galaxies are just markers or test particles scattered through space, which indicate how the material context of the universe is distributed.

Galaxies are clustered. Some are in small groups such as our own Local Group, of which the Milky Way and the Andromeda Galaxy are the dominant members. Others are in big clusters with hundreds of members. But on the really large scale, the universe genuinely does seem simple and smooth. If one imagined a box whose sides were one hundred million light-years (dimensions still small compared to the observable universe) its contents would look about the same wherever we placed it. In other words there is a well-defined sense in which the observable universe is indeed roughly homogeneous. The brightest million galaxies are fairly uniform over the sky, and as we look at still fainter galaxies, probing to greater distances, clustering becomes less evident and the sky appears smoother. We are not in the kind of universe discussed by Carl Charlier in the early 1900s (Figure 8), containing clusters of clusters of clusters *ad infinitum*. Unless we are anti-

Copernican and assign ourselves a privileged central position, this apparent isotropy implies that the universe is roughly isotropic about any galaxy – that the universe is homogeneous, and all parts evolved in the same way and had the same history.

'The fox knows many things but the hedgehog knows one big thing.' And cosmologists are, in this sense, perhaps, the most hedgehog-like of all scientists. Their subject boasts few firm facts, but each with great ramifications. The first key fact emerged in 1929, when Edwin Hubble enunciated his famous law, that galaxies recede from us at speeds proportional to their distance. We seem to inhabit a homogeneous universe where the distances between any widely separated pair of galaxies stretches as some uniform function of time. This does not imply that we are in a special 'plague spot'. Hypothetical observers on any other galaxy would see a similar isotropic expansion away from them.

Hubble's work suggested that galaxies would have been crowded together in the past, and that there was some kind of 'beginning' ten or twenty billion years ago. But he had no direct evidence for cosmic evolution. Indeed, the steady-state theory, proposed in 1948, envisaged continuous creation of new matter and new galaxies, so that despite the expansion the overall cosmic scene never changed.

Origins in a hot 'big bang'

We would not expect any cosmic evolutionary trend unless we can probe at least several billion years back in time. This entails studying objects billions of light-years away, which may be invisibly faint even with the largest telescopes. It was Martin Ryle and his colleagues, in Cambridge in the late 1950s, who found the first evidence that our entire universe was evolving. His telescope could pick up radio waves from some active galaxies (the ones that we now think harbour massive black holes), even when these were too far away to be observed optically. He could not determine a red-shift or distance from radio measurements alone, but assumed that, statistically at least, the ones appearing faint were more distant than those appearing intense. He counted the numbers with various apparent intensities, and found that there were too many apparently faint ones (in other words, those at large distances) compared to brighter and closer ones. This was discomforting to the 'steady statesmen', but compatible with an evolving universe if galaxies were more prone to violent outbursts in the remote past, when they were young.

But the clinching evidence for a so-called 'big bang' came in 1965, when Arno Penzias and Robert Wilson at the Bell Telephone Laboratories detected the cosmic microwave background radiation. This discovery was accidental: their prime motive was a practical one – communication with artificial satellites – and they did not immediately realise what they had found. But the excess background noise in their instruments, which they could neither eliminate nor account for, meant that even intergalactic space is not completely cold: it is about three degrees above absolute zero. This may not sound much, but it implies that there are about 400 million quanta of radiation (photons) per cubic metre. Indeed, there are 10^8 photons for each atom in the universe (Figure 9).

There is no way of accounting for this radiation, and its spectrum and isotropy, except on the hypothesis that it is a relic of a phase when the entire universe was hot, dense and opaque. It seems that everything in the universe once constituted an exceedingly com-

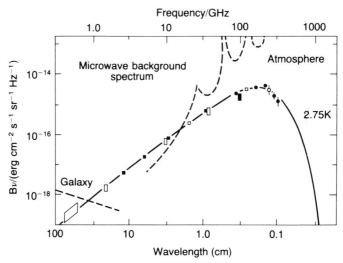

9. *The spectrum of the microwave background radiation is best fit by a black body with a temperature of 2.67K above absolute gas. Measurements at frequencies near the peak of the spectrum must be made from above the atmosphere (or indirectly via studies of the excitation of interstellar molecules); at lower frequencies the radiation can be detected at ground level. Penzias and Wilson made their original measurement (marked on the diagram) at about 4 GHz. The temperature is uniform over the sky with a precision of one part in 10^4 – this is the best evidence that the overall cosmic expansion is isotropic, and that the universe is highly homogeneous on the largest scales we can observe (billions of light-years).*

pressed and hot gas, hotter than the centres of stars. The intense radiation in this compressed fireball, though cooled and diluted by the subsequent expansion (the wavelengths being stretched and red-shifted), would still be around. The universe would have become transparent after about a million years, when the temperature fell to 3000 degrees and hydrogen recombined. The photons that Penzias and Wilson detected have travelled uninterruptedly since that time – in other words for about 99.99% of cosmic history (Figure 11).

Is there any corroboration of the primordial 'fireball'? According to this concept, when the universe was only a few minutes old, it would have been at a billion degrees. Nuclear reactions would have occurred as it cooled through this temperature range. These can be calculated in detail. The material emerging from the fireball would be about 75% hydrogen and 25% helium. This is specially gratifying, because the theory of synthesis of elements in stars and supernovae, which works so well for carbon, iron, etc., was always hard pressed to explain why there was so much helium, and why the helium was so uniform in its abundance. Attributing helium formation to the big bang thus solved a long-standing problem in nucleogenesis, and bolstered cosmologists' confidence in extrapolating right back to the first few seconds of the universe's history, and assuming that the laws of microphysics were the same as now. (Without this firm link with local physics, cosmology risks degenerating into *ad hoc* explanations on the level of 'just so' stories.)

More detailed calculations, combined with better observations of background radiation and of element abundances, have strengthened the consensus that the hot big bang model

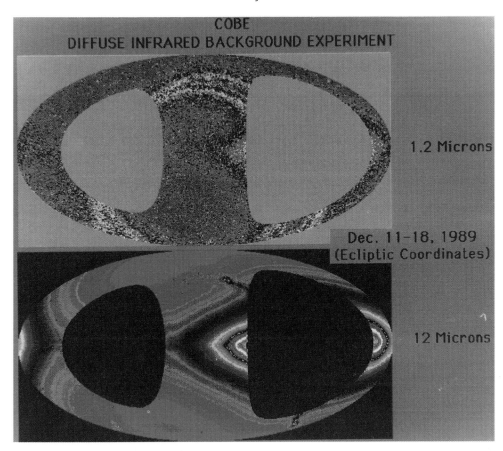

10. *The Cosmic Background Explorer (COBE) collected the data for these sky maps. Different colours denote regions of different sky brightness.*

is basically valid. It is not yet firm dogma. But the hot big bang model certainly seems more plausible than any equally specific alternative – I would personally make the stronger claim that it has more than an even chance of survival.

Consequently, most of us adopt a cosmogonic framework like this. Stars and galaxies all emerged from a universal 'thermal soup', at a temperature of ten billion degrees, expanding on a time scale of one second. It was initially almost smooth and featureless, but not quite: there were fluctuations from place to place in the density or expansion rate. Embryonic galaxies – slightly overdense regions whose expanion rate lagged behind the mean value – evolved into disjoint clouds whose internal expansion eventually halted. These protogalactic clouds collapsed to form galaxies when the universe was perhaps 10% of its present age; subsequently, the galaxies grouped into gravitationally bound clusters. This latter process can be well simulated by *N*-body dynamical computer calculations. Sverne Aarseth made a movie which illustrates cluster formation. This movie shows the last 90% of cosmic history, speeded up by a factor of about 10^{16}: if the galaxies start off distributed almost uniformly, but with random fluctuations, then regions

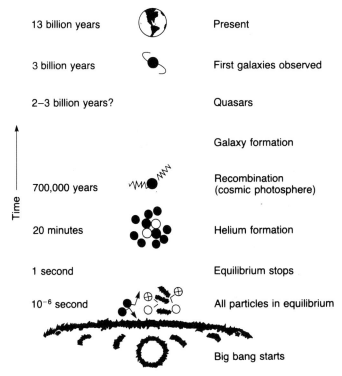

13 billion years — Present

3 billion years — First galaxies observed

2–3 billion years? — Quasars

Galaxy formation

700,000 years — Recombination (cosmic photosphere)

20 minutes — Helium formation

1 second — Equilibrium stops

10^{-6} second — All particles in equilibrium

Big bang starts

Time

11. *The history of the universe according to the so-called 'hot big bang' theory.*

containing a slightly above-average concentration of galaxies condense into gravitationally bound groups and clusters with a gratifying resemblance to the real groupings seen in our actual sky.

Gravity is crucial here, as it is for the internal evolution within each galaxy. Gravity drives things *further from equilibrium*. When gravitating systems *lose* energy they get *hotter*. A homely instance of this is the way an artificial satellite speeds up as it spirals downward due to atmospheric drag. Another example is offered by Kelvin's ideas on how the Sun would evolve if its radiative losses were not compensated for by the energy from nuclear fusion. A star that loses energy and deflates ends up with a hotter centre than before: to establish a new and more compact equilibrium where pressure can balance a (now stronger) gravitational force, the central temperature must rise. This requires an increase in the star's internal thermal energy, and the star cannot actually radiate more than half the gravitational energy released. This apparent 'antithermodynamic' behaviour amplifies density contrasts and creates temperature gradients, a prerequisite for the evolution of any complexity. A one-sentence answer to the question: 'What is happening in the universe?' might go like this: 'Gravitational binding energy is being released as stars, galaxies and clusters progressively contract, this inexorable trend being delayed by rotation, nuclear energy, and the sheer scale of astronomical systems, which makes things happen slowly and stave off gravity's final victory.'

Martin J. Rees

The second key feature of gravity is its *weakness*. The gravitational force within an atom is about forty powers of ten weaker than the electric force that binds it. But everything has the same sign of 'gravitational charge': it is a long-range force with no cancellation. So gravity holds sway on sufficiently large scales, scales which are vast *because* gravity is so weak. If gravitation were somewhat stronger – thirty and not forty powers of ten weaker than microphysical forces, for instance – a small-scale speeded-up universe could exist, in which gravitationally bound fusion reactions had 10^{-15} of the Sun's mass, and lived for less than a year. This might not allow enough time for complex systems to evolve. There would be fewer powers of ten between astrophysical time scales and the basic microscopic time-scales for physical or chemical reactions. Complex structures, moreover, could not become very large without themselves being crushed by gravity. Our universe is vast and diffuse, and evolves so slowly, *because* gravity is so weak. Its extravagant scale, billions of light-years, is necessary to provide enough time for the cooling of elements inside stars, and for interesting complexity to evolve around even one star in just one galaxy.

One stumbling-block in understanding galaxies, incidentally, is the rather embarrassing fact that 90% of their mass is unaccounted for. When we study motions in the outer parts of galaxies, and the relative motions of galaxies gravitationally bound in groups or clusters, we infer that the galaxies are feeling the gravitational pull of ten times more stuff than we see (Figure 12). There is no reason to be amazed by this – no reason why everything in the universe should shine conspicuously. What are the candidates for this 'dark matter'? It could be in very faint stars too small to have ignited their nuclear fuel, or, alternatively, in the remnants of massive stars which were bright in the early phases of galactic history but have now all died out. A third idea, much discussed in recent years, is that the primordial fireball might have had extra ingredients apart from the ordinary atoms and radiation that we observe, and that elementary particles of some novel type could collectively exert large-scale gravitational forces.

There are various observational ways of deciding between such varied options. But it would be of special interest to particle physicists if astronomers were to learn more about neutrinos – ghostly and elusive particles which hardly interact at all with ordinary matter – or discover some fundamentally new particles (for instance, the photinos whose existence some theorists predict). We would then have to view the galaxies, the stars, and ourselves in a downgraded perspective. Over four centuries ago, Copernicus dethroned the Earth from a central position. Early this century, Harlow Shapley and Hubble demoted us from any privileged location in space. But now even 'particle chauvinism' may have to be abandoned. The protons, nuclei and electrons, which we and the entire astronomical world are made up of, could be a kind of afterthought in a cosmos where neutrinos or photinos control the overall dynamics.

Dark matter – what it is, and how much there is of it – is relevant to cosmogony, especially to the details of galaxy formation. But it is even more crucial for the long-term future – for eschatology. Will our universe expand for ever, and the galaxies fade and disperse into an ultimate heat death? Or will it recollapse, so that our descendants share the fate of an astronaut who falls into a black hole, the firmament falling on their heads to recreate a fireball like that from which it emerged?

229

12. *A disc galaxy viewed edge on (NGC 4565). Studies of the rotation speed of gas in the outlying parts of such discs suggest that this gas is 'feeling' the gravitational pull of more material than the stars and gas we actually see. Luminous galaxies are apparently embedded in a more massive and extensive cloud of 'dark' matter. Corroboration of the existence of 'dark' matter surrounding galaxies comes from analyses of the motions of gravitationally bound groups and clusters of galaxies: the total masses inferred from the relative motions of galaxies are about ten times higher than those inferred from the internal dynamics of their luminous cores.*

Imagine a large sphere or asteroid, which is shattered by an explosion, the debris flying off in all directions. Each fragment feels the gravitational pull of all the others, causing deceleration. If the explosion were sufficiently violent, then the debris would fly apart for ever. But if the fragments were not moving quite so fast, gravity might bind them together strongly enough to bring the expansion to a halt. The material would then fall together again. According to general relativity, this same argument holds for the universe. In the case of the galaxies, which for the purpose of this argument are envisaged as 'fragments' of the expanding universe, we know the expansion velocity. What we do not know so well is the amount of gravitating matter tending to brake the expansion. It is easy to calculate how much will be needed to bring it to a halt: it works out at about three atoms per cubic metre. Were the average concentration below this 'critical' density we would expect the universe to continue expanding for ever, but if the mean density exceeded this value, a big crunch would seem inevitable.

Even if we include the dynamically inferred dark matter in galaxies and clusters, the mean density still falls short of the critical value by a factor of about five. But there could be some even more elusive material between clusters of galaxies. Until our knowledge of dark matter candidates is less biased and more complete, we will not have a reliable long-term forecast for the universe.

The ultra-early universe: initial conditions

Although the alternative long-range futures seem very different, Figure 13 highlights a puzzle. The initial conditions that could have led to anything like our present universe are actually very restrictive, compared to the range of possibilities that might have been set up. We know that our universe is still expanding after 10^{10} years. Had it recollapsed sooner, there would have been no time for stars to evolve. If it had collapsed after less than a million years, it would have remained opaque, precluding thermodynamic disequilibrium. The expansion rate cannot, however, be too much faster than parabolic,

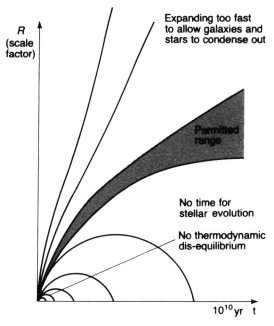

13. This diagram depicts the 'scale factor' for a family of hypothetical isotropic universes which all start off expanding, but are decelerated by the gravitational attraction excited by each part on all other parts. We do not know whether our universe will go on expanding for ever. We do, however, know that our universe is still expanding, after 10 billion years, but that it is not expanding so rapidly that galaxies could not condense out to form gravitationally bound systems. There is such a sense in which the initial conditions seem to have been 'finely tuned'. Moreover, the initial conditions seem even more special when we note that more general cosmological models (e.g. those which are anisotropic) have other degrees of freedom open to them. Recent ideas in high energy physics allow us for the first time to approach this question of initial conditions scientifically.

otherwise the expansion kinetic energy would have overwhelmed gravity, and the clouds that developed into galaxies would never have been able to condense out. (This is equivalent to saying that the present density is not orders of magnitude below the critical density.) There is, therefore, a sense in which the dynamics of the early universe must have been finely tuned. In Newtonian terms the fractional difference between the initial potential and kinetic energies of any spherical region must have been very small.

So why was the universe set up expanding in this special way? And there are other issues that similarly baffle us. Why does the universe contain fluctuation, while being so homogeneous overall? Why are there 10^8 photons for each particle?

We have pushed the chain of inference right back from the early solar system to the cosmic fireball at $t = 1$ s. But, conceptually, we are in little better shape. We still seem to be appealing to initial conditions: 'Things are as they are because they were as they were.' Our inferences come up against a barrier, just as did the ancient Indian cosmologists who envisaged the Earth supported by four elephants standing on a giant turtle, but did not know what held the turtle up.

Maybe key features of the universe were imprinted at still earlier stages than $t = 1$ s. The further back one extrapolates, the less confidence one has in the adequacy or applicability of known physics. For instance, the material would exceed nuclear densities for the first microsecond. But if one thinks of time on a logarithmic scale, to ignore these early eras is a severe omission indeed. Theorists differ in how far they are prepared to extrapolate back with a straight face. Some have higher credulity thresholds than others. But those whose intellectual habitat is the 'gee whiz' fringe of particle physics are interested in the possibility that the early universe may once have been at colossally high temperatures. To motivate this interest, let us digress for a moment to discuss the basic physical forces.

These forces are just four in number: electromagnetism, the weak force (important for radiative decay and neutrinos), the strong or nuclear force, and gravity. Physicists would like to discover some inter-relation between them, to interpret them as different manifestations of a single primeval force. The first modern step towards this unification was the Salam–Weinberg theory of the late 1960s, relating the electromagnetic and the weak forces. The basic idea is that at high energies these two forces are the same. They acquire distinctive identities only below some critical energy. Energies in particle physics are measured in giga electron volts, GeV for short, and the critical energy for this unification is about 100 GeV. This energy can just be reached by large particle accelerators, and the Salam–Weinberg theory has been vindicated by experiments at CERN. This development may prove as important in its way as Clark Maxwell's achievement one hundred years ago in showing electrical and magnetic effects to be manifestations of a single underlying force (Figure 14).

The next goal is to unify the electro-weak force with the strong or nuclear forces – to develop a so-called grand unified theory (GUT) of all the forces governing the microphysical world. But a stumbling-block here is that the critical energy at which the so-called symmetry breaking occurs, the energy which is 100 GeV for the Salam-Weinberg theory, is thought to be 10^{15} GeV for the grand unification. This is a million million times higher than any feasible experiments can reach. It is hard, therefore, to test these theories

Martin J. Rees

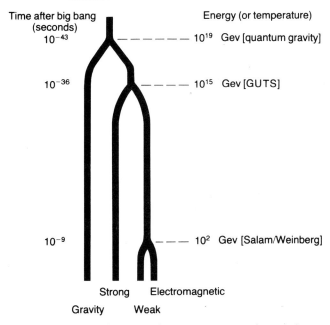

Unification scheme for the forces

Time after big bang
(seconds)

Energy (or temperature)

10^{-43} — — — — — — 10^{19} Gev [quantum gravity]

10^{-36} — — — — 10^{15} Gev [GUTS]

10^{-9} — — — 10^{2} Gev [Salam/Weinberg]

Strong   Electromagnetic

Gravity   Weak

14. *Schematic diagram showing how, according to some current theories, the four forces of nature are unified at high temperatures which have been attained in the initial instants of the cosmic expansion.*

on Earth. Only tiny effects are predicted in our low-energy world: for instance, protons would decay very slowly. But if we are emboldened to extrapolate the big bang theory far enough we find that in the first 10^{-36} s, but only then, the particles would be so energetic that they would all be colliding at 10^{15} GeV. So perhaps the early universe was the only accelerator where the requisite energy for unifying the forces could ever be reached. However, this accelerator shut down ten billion years ago, so one can learn nothing from its activities unless the 10^{-36} s era left some fossils behind, just as most of the helium in the universe was left behind from the first few minutes. Physicists would seize enthusiastically at even the most trifling vestige surviving from that phase. But it has left very conspicuous traces indeed: it may be that all the atoms in the universe are essentially a fossil from 10^{-36} s.

If you were setting up a universe in the simplest way, you might make it symmetrical between matter and antimatter, preparing it with equal numbers of protons and antiprotons. But the particles and antiparticles would then all annihilate as the universe expanded and cooled. We would end up with black-body radiation but no matter, no atoms, and no galaxies. However, grand unified theories evade this problem in a way first outlined in a prescient paper by Andrei Sakharov, written in 1967. Although these theories predict that proton decay is incredibly slow now, at 10^{-36} s protons could readily be created or destroyed. As Sakharov first realised, the expansion and symmetry breaking, according

233

to these theories, introduce a slight but calculable favouritism for creation of particles rather than their antiparticles. So that for every 10^8 proton–antiproton pairs, there is one extra proton. As the universe cools, antiprotons all annihilate with protons, giving photons. But for every 10^8 photons thereby produced, there is one proton that survives, because it cannot find a mate to annihilate with. The photons, now cooled to very low energies, constitute the 3K background. There are indeed $\approx 10^8$ of them for every atom. So the entire matter content of the universe could result from a small fractional bias in favour of matter over antimatter, imposed as the universe first cooled below 10^{15} GeV.

Grand unified theories are still tentative, but they at least bring a new set of questions – the origin of matter, for instance – within the scope of serious discussion. The realisation that protons are not strictly conserved suggests, moreover, that the universe may possess no conserved quantities other than those, e.g. total electric charge, which are strictly zero. This, combined with the concept of a so-called 'inflationary' phase, whereby our universe could have originated from even a quantum fluctuation, opens the way to envisaging an almost *ex nihilo* origin for our entire universe.

Frontiers and limits

At the start of this chapter we looked at ordinary stars where scientists feel fairly confident that they know the relevant physics. When conditions are more extreme, such as in galactic nuclei, we are less confident, though it is astonishing how far we can go without running up against a contradiction. One theme that has emerged is the interdependence of different phenomena. The everyday world is determined by atomic structure, stars are determined by the physics of atomic nuclei, and the much larger structures (galaxies and clusters) may be gravitationally bound only because they are embedded in clouds of subatomic particles which are relics of a high-energy phase.

But in considering the early big bang, or gravitation collapse inside black holes, we are confronted by conditions so extreme that we know for sure that we do *not* know enough physics. In particular, physics is incomplete and conceptually unsatisfactory in that we lack an adequate theory of *quantum gravity*. The two great foundations of twentieth-century physics are the quantum principle and Einstein's general relativity. The theoretical super-structures erected on these foundations are still disjoint: there is generally no overlap between their respective domains of relevance. Quantum effects are crucial on the microscopic level of the single elementary particle, but gravitational forces between individual particles are negligible. Gravitational effects are manifested only on the scale of planets, stars and galaxies, where quantum effects and the uncertainty principle can be ignored. But when the universe was squeezed to colossal densities and temperatures, gravity could be important on the scale of a single particle, a single thermal quantum. This happens at 10^{-43} s, the Planck time. The effects of quantum gravity would then be dominant, and even the boldest physicists can extrapolate back no further.

Despite the difficulties, some theorists believe it is no longer premature to explore what physical laws prevailed at the Planck time, and have already come up with fascinating ideas; there is no consensus, though, about which concepts might really 'fly'. We must certainly jettison cherished common-sense notions of space and time: space-time on this tiny scale may have a chaotic foam-like structure, with no well-defined arrow of time, and there may be no time-like dimension at all. A second idea, generating euphoric optimism at the moment, is that on the tiniest scale space may have extra dimensions. These extra dimensions are not manifest in the everyday world because they are compactified, rather as a sheet of paper, a two-dimensional surface, might look like a one-dimensional line if rolled up very tightly.

In the light of these speculations about the beginning of time, the demarcation between initial conditions and the laws governing time-development become blurred. Maybe there is only one self-consistent way the universe could have started off. Our ignorance of the physics governing the first microsecond of cosmic expansion has some analogy with the way ignorance of atomic structure stymied nineteenth-century speculations about the Sun. But there is an important difference. Atoms and nuclei could be probed in the laboratory, whereas the early universe is the only place manifesting these ultra-high-energy phenomena. This makes it hard to test unified theories, but at least offers cosmologists a relationship with their physicist colleagues that is symbiotic rather than parasitic.

What is the real chance of clarifying these fundamental questions about physical reality in the next few decades? The prospects are not necessarily hopeless; it is not presumptuous to try. It is, after all, complexity not sheer size that makes a process hard to comprehend. For example, we already understand the inside of the Sun better than the interior of the Earth. The Earth is more difficult to understand because the temperatures and pressures inside it are less extreme than in the Sun, where it is so hot that everything is broken down into its constituent atoms. For analogous reasons it is harder to understand the tiniest living organism than any large-scale inanimate phenomenon. In the earliest stages of the primordial fireball, matter would surely be reduced, broken down, to its most primitive constituents. So maybe we can realistically hope to understand why the universe is expanding the way it is, and thereby inter-relate the cosmic and microphysical scales in a more profound way. If this goal were achieved, it would supremely exemplify what the physicist Eugene Wigner called the 'unreasonable effectiveness of mathematics in the physical sciences'. It would mean, in a sense, the end of fundamental physics. But this would emphatically not mean the end of challenging science.

A metaphor for what the physicist does is this: suppose you were unfamiliar with the game of chess. Then, just by watching games being played, you could infer what the rules were. The physicist, likewise, finds patterns in the natural world and learns what dynamics and transformations govern its basic elements. In chess, learning how the pieces move is just a trivial preliminary to the absorbing progression from novice to Grand Master. The whole point and interest of that game lies in exploring the complexity implicit in a few deceptively simple rules. Likewise, all that has happened in the universe over the last ten billion years – the formation of galaxies and the intricate evolution on a planet around at least one star that has led to creatures able to wonder about it all – may be implicit in a few fundamental equations of physics, but exploring all this offers an unending quest and challenge that has barely begun.

Acknowledgements

Abbreviations

AAT – Anglo-Australian Telescope Board; ESO – European Southern Observatory; JPL – Jet Propulsion Laboratory, USA; NASA – National Aeronautics and Space Administration, USA; NOAO – National Optical Administration Observatories, USA; RAS – Royal Astronomical Society; RGO – Royal Greenwich Observatory, UK; ROE – Royal Observatory Edinburgh, UK

1.1 RAS; 1.2 ROE; 1.3 RGO; 1.4 Ann Ronan Picture Library; 1.5 Paul Dogherty; 1.6 Van Speybroeck *et al.*, *Astrophys. J. Lett.*, **234**, L45; 1.7 Novosti Press Agency; 1.8, 1.9 NASA; 1.10 *Nature*, 1914; 1.11, 1.12 RAS

2.1 *J. Liverpool Astronomical Society* (1887–88), **6**, 118; 2.2–2.7 NASA; 2.8 Harold Hill; 2.9 Ron Arbour; 2.10 Lowell Observatory Photograph; 2.11–2.13 NASA; 2.14 Dr Philip James, University of Toledo, NASA

3.1, 3.2 John Rogers; 3.3 NASA; 3.4, 3.6 Dr Bradford, A. Smith and the National Space Science Data Center through the World Data Center A for Rockets and Satellites; 3.5 John Rogers; 3.7 NASA; 3.8, 3.9 JPL

4.1 Richard McKim; 4.2–4.7 JPL; 4.8 Paul Dogherty; 4.9 JPL

5.1–5.3 JPL; 5.4 NASA; 5.5–5.9 JPL; 5.10 Clyde Tombaugh, March 1930

6.1 David W. Hughes; 6.2 ROE; 6.3 1986 Max-Planck-Institut Für Aeronomie, courtesy Dr H. U. Keller; 6.5–6.8 David W. Hughes; 6.9 ROE

7.1, 7.2 Paul Dogherty; 7.4 Jean Dragesco; 7.5 NASA; 7.6 Big Bear Solar Observatory; 7.7 Ian Turton; 7.8 H. J. P. Arnold/Space Frontiers Ltd; 7.9 NASA; 7.10 Armagh Planetarium

8.1 Dr Bradford, A. Smith and the National Space Science Data Center through the World Data Center A for Rockets and Satellites; 8.2–8.6 ROE; 8.7 Akira Fujii; 8.8 AAT

236

Acknowledgements

9.1 Ron Arbour; 9.2 *BAAJ*, June 1979; 9.3–9.8 after *Astronomy Now*; 9.9, 9.10 Paintings by Steven Simpson, Courtesy of *Sky & Telescope*; 9.11 Space Telescope Science Institute

10.1 1979 ESO; 10.2–10.4 RGO/AAT; 10.5 P. Murdin; 10.6, 10.10 RGO; 10.11 ROE

11.2 ROE & AAT; 11.3 RAS; 11.4 NOAO

12.1 National Gallery Picture Library; 12.2 Akira Fujii; 12.3 Paul Dogherty; 12.4 Akira Fujii; 12.5 NASA; 12.6 Rutherford Appleton Laboratory; 12.7 Julian Baum; 12.8 AAT

13.1 RGO; 13.2 H. C. Arp, in *Optical Jets in Galaxies*, ed. B. Battrick and J. Mort, ESA Publications, 1981; 13.3 ESO; 13.4 K. Pounds and The European Space Agency (1987); 13.5 A. Wandel & R. F. Mushotzky *Astrophys. J.* (1986), **306**, L63–4; 13.6 from *The New Physics*, ed. Paul Davies, p. 173, Cambridge University Press, 1989; 13.7 from C. M. Gaskell, *Astrophys. J.* (1982), **252**, 447; 13.8 from M. Mackan and W. L. W. Sargent, *Astrophys. J.* (1982), **254**, 33; 13.9 from R. A. Perley, J. W. Dreher and J. J. Cawan, *Astrophys. J.* 1984), **285**, L35; 13.10 from *The New Physics*, ed. Paul Davies, p. 179; 13.11 from T. J. Pearson *et al.*, *Extragalactic Radio Sources*, ed. D. S. Heeschen and C. M. Wade, p. 356, D. Reidel, 1982; 13.12 from *The New Physics*, ed. Paul Davies; 13.13 from M. J. Rees, *The Physics of Non-Thermal Radio Sources*, ed. G. Setti, p. 107, D. Reidel, 1976

14.1 CERN; 14.2 from *The New Physics*, ed. Paul. Davies; 14.7 CERN

15.1 David Malin, AAT; 15.2 Akira Fujii; 15.3, 15.4 from *Origins*, ed. A. C. Fabian, Cambridge University Press, 1989; 15.6 AAT; 15.7 Dept of Astrophysics, University of Oxford; 15.8, 15.9, 15.11, 15.13, 15.14 from *Origins*, ed. A. C. Fabian